Stand-Off Detection of Suicide Bombers and Mobile Subjects

NATO Security through Science Series

This Series presents the results of scientific meetings supported under the NATO Programme for Security through Science (STS)

Meetings supported by the NATO STS Programme are in security-related priority areas of Defence Against Terrorism or Countering Other Threats to Security. The types of meeting supported are generally "Advanced Study Institute" and "Advanced Research Workshops". The NATO STS Series collects together the results of these meetings. The meetings are co-organized by scientist from NATO countries and scientists from NATO's "Partner" or "Mediterranean Dialogue" countries. The observations and recommendations made at the meetings, as well as the contents of the volumes in the Series, reflect those of participants and contributors only; they should not necessarily be regarded as reflecting NATO views or policy.

Advanced Study Institutes (ASI) are high-level tutorial courses to convey the latest developments in a subject to an advanced-level audience

Advanced Research Workshops (ARW) are expert meetings where an intense but informal exchange of views at the frontiers of a subject aims at identifying directions for future actions

Following a transformation of the programme in 2004 the Series has been re-named and re-organised. Recent volumes on topics not related to security, which result from meetings supported under the programme earlier, may be found in the NATO Science Series.

The Series is published by IOS Press, Amsterdam, and Springer, Dordrecht, in conjunction with the NATO Public Diplomacy Division.

Sub-Series

A. Chemistry and Biology	Springer
B. Physics and Biophysics	Springer
C. Environmental Security	Springer
D. Information and Communication Security	IOS Press
E. Human and Societal Dynamics	IOS Press

http://www.nato.int/science
http://www.springer.com
http://www.iospress.nl

Series B: Physics and Biophysics

Stand-Off Detection of Suicide Bombers and Mobile Subjects

edited by

Hiltmar Schubert
Fraunhofer-Institute for Chemical Technology (ICT),
Pfinztal, Germany

and

A. Rimski-Korsakov
V.G. Khlopin Radium Institute (KRI),
St. Petersburg, Russia

Springer

Published in cooperation with NATO Public Diplomacy Division

Proceedings of the NATO Advanced Research Workshop on
Stand-Off Detection of Suicide Bombers and Mobile Subjects
Pfinztal, Germany
13–14 December 2005

A C.I.P. Catalogue record for this book is available from the Library of Congress.

ISBN-10 1-4020-5158-1 (PB)
ISBN-13 978-1-4020-5158-6 (PB)
ISBN-10 1-4020-5157-3 (HB)
ISBN-13 978-1-4020-5157-9 (HB)
ISBN-10 1-4020-5159-X (e-book)
ISBN-13 978-1-4020-5159-3 (e-book)

Published by Springer,
P.O. Box 17, 3300 AA Dordrecht, The Netherlands.

www.springer.com

Printed on acid-free paper

CONTENTS

PREFACE

These proceedings contain the presentations of a NATO-Russian Advanced Research Workshop (ARW) held at the Fraunhofer Institute for Chemical Technology (ICT) near Karlsruhe, Germany, attended also by most of the members of the "Explosive Detection Working Group" (EDWG) of the "Advisory Panel on Security-Related Civil Science and Technology".

The ARW, approved by the NATO Science Committee, is the fifth NATO-Workshop during the last 3 years dealing with the detection of explosives in connection with counter-terrorism.

The special goal of this workshop was the exchange of knowledge and experiences in the special field of "Stand-Off Detection" and to fix the "State of the Art" in this field of research.

At the end of the workshop proposals of Research & Development were presented (but not included in the proceedings), which were evaluated by the present members of the EDWG during the following day of the workshop.

One of the first measures to counter the threat will be to stop the transport of illicit materials, that means in our case to detect explosive charges before they are initiated by terrorists.

These detection methods have to be in a secret and also harmless way and in an appropriate distance. Under consideration of scientific applied possibilities a range of a minimum of 1–10 m should be aimed in a first step. Larger distances in a mid or long range are welcome and "Out-door situations" are taken into account. Different technologies in fused systems should also be considered in later steps of applied research as a necessary approach.

The presentations contain methods of Bulk Detection and of Vapour/Particle Detection concerning electromagnetic approaches on the one side and remote trace techniques on the other side.

In the field of electromagnetic methods, technologies in the frequency range of 10–100 GHz and 300–3000 GHz were of special interest, but also millimetre and terahertz frequencies and dielectric material measurements were of significance.

Belonging to laser and vapour/particle detection, a number of presentations contain remote sampling laser techniques and vapour/particle sensing. Methods based on stand-off laser techniques were recommended.

New MIP-based low-cost sensors, semiconductor sensors and spectroscopical sensors (Raman, FTIR, NQR, etc) were presented.

Cochairmen
Hiltmar Schubert
Alexander Rimski-Korskov

ACKNOWLEDGEMENT

It was a pleasure for me to organize this ARW together with Prof. Dr. Rimski-Korsakov, Director of the Khlopin Instutut, St. Petersburg, Russia, and would like to thank him for selecting and motivating qualified high rank Russian expert in the field of Stand-Off Detection of Explosives or Devices.

I thank the directors of the Fraunhofer Institut for Chemical Technology (ICT), Prof. Dr. Peter Eyerer and Dr. Peter Elsner and their coworkers for their support.

It is owing to the organization staff of ICT and to my secretary, Mrs. Birgit Baumgartner, that preparation and passing off the workshop result in an excellent style.

The workshop causes a large interest of the members of the international "Explosives Detection Working Group" of the "Advisory Panel on Security-Related Civil Science & Technology" and its Programme Director Prof. F.C. Rodrigues. I thank them all for their support, interest and cooperation including all participants of the workshop from east and west.

Hiltmar Schubert

1. PROBLEM OF EXPLOSIVE DETECTION ON HUMAN SUBJECTS

A. Rimski-Korsakov[1]
V.G. Khlopin Radium Institute, Saint-Petersburg, Russia

Abstract. Requirements of the NATO-Russian Explosive Detection Working Group (EDWG) of the NATO Advisory Panel on Security-Related Civil Science and Technology concerning systems for standoff detection of explosives carried by mobile subjects are outlined. Prospective technologies are listed. Potential challenges facing the future detection systems in real-life environments are briefly discussed.

In 2004–2005 the NATO-Russian Explosive Detection Working Group (EDWG) of the NATO Advisory Panel on Security-Related Civil Science and Technology prepared a proposal for a special research program to develop methods and equipment suitable for standoff detection of explosives carried by mobile subjects (primarily—suicide bombers). This problem has been ascribed high priority because of absence of any applicable technology to deal with the multiple incidents involving suicide bombers and car bombs, used recently by terrorists both in NATO countries and Russia. This Advanced Research Workshop aims to evaluate possible methods and technologies to solve this problem as soon as possible by developing first a laboratory demonstration prototype of such equipment and then introducing the technology as commercial product for field use.

The EDWG has formulated mission scenarios and requirements for technology, which could be acceptable to solve the problem:

- Detection should be harmless and unnoticeable to person under inspection;
- Detection device may be portable or stationary;
- The sensor (or sensors) may be separate from the registration unit;
- The detection distance should be between 1 and 10 m;
- The HEX carrier person will be moving, so detection must be done in "real time" — preferably in times less than one second.

[1] E-mail: ark@atom.nw.ru

H. Schubert and A. Rimski-Korsakov (eds.), Stand-off Detection of Suicide Bombers and Mobile Subjects, 1–3.

Regarding these requirements various existing detection principles have been discussed and broken into two groups—bulk detection methods and trace detection methods:
Bulk detection:

- Terahertz-Wave "Vision";
- MM-Wave and Microwave imaging;
- NQR investigation;
- Advanced (low-dose) X-ray Imaging.

Trace detection (distributed detector network):

- Chemical Detectors based on molecular fluorescence;
- Ion Mobility Spectroscopy;
- Gas chromatography with surface acoustic wave detector;
- Standoff laser techniques.

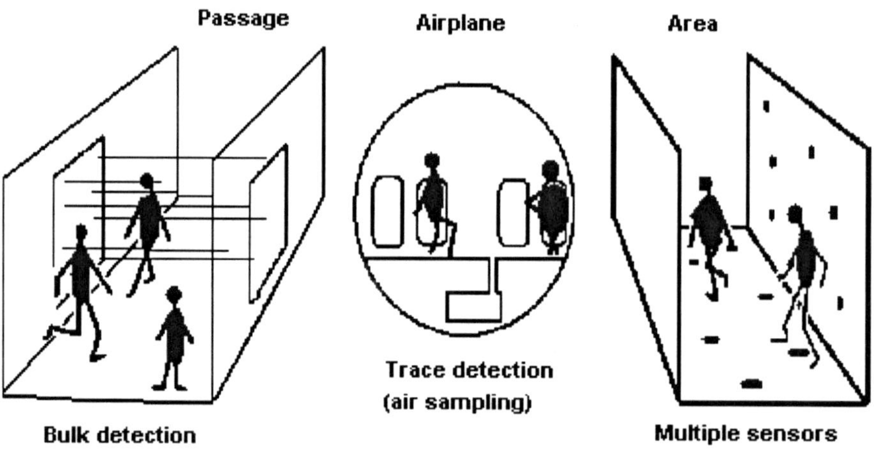

Important issue of possible detector positioning has to take into account a compromise between detection range (not less than 1 m) and the range at which the moving subject will not notice the inspection. This requirement excludes usual "portal" technology, if the subject has to stop or to pass through a narrow gate, but permits to consider real-time inspection of people moving in some wide passageway. Various trace detectors could be used also in a "sealed" closed room (e.g. passenger cabin of an airliner) — in this case air can be sampled by air condition unit and some warning signal of presence of explosive generated for security action (e.g. to prevent take-off of a flight). Similar warning signal could be generated if traces of explosive are detected by a network of chemical sensors in a larger area, such as shopping mall, stadium or transport terminal.

Whatever method or technology will be proposed, the application of the method will meet with many difficulties and restrictions, both obvious and not so obvious.

Obvious difficulties, for instance would require negligible dosage (if any), if ionizing radiation (such as X-rays) is used. Another obvious requirement is very short detection time (in range of seconds), since the person under inspection will be free to move. It is also obvious, that false alarm rate must be very low, because the method will be applied to many people, where even 1% of false alarms will make operation practically impossible.

There are more difficulties, which are not so obvious—but may prove very hard to overcome in real application of any standoff technology. For instance, any covert imaging of people may be unacceptable to public opinion, so that images generated will have to be made distorted or symbolic (but can be interpreted nevertheless). Another difficulty can arise from the fact that person carrying the explosive must be "tagged" somehow, and this must be done without his/her knowledge, so that appropriate security action can be done on a right person. For such "tagging" the method has very little time, since subject is moving and will go out of inspection area in seconds. One of the most serious difficulties for standoff inspection is the fact that human operators cannot work efficiently for long periods of time. Their attention to the job will be impaired in several hours, and the efficiency of detection may be lost unless they have some prompt from automatic detection system. This can be avoided only if some autonomous computer analysis of detector signals (without human intervention) is added to any technology that will be used in the real field circumstances. In such case the computer will send alarm in all "suspicious" cases, and human operator will make decision (maybe using some additional data) whether security enforcement personnel should handle this case.

All these considerations must be taken into account when various technology proposals will be discussed in our workshop. Of course, some of the problems outlined above can be solved if we use several methods in combination, which usually increases selectivity and reliability of detection and strongly reduces false alarm rate.

2. MS-SRIP—MICROWAVE SYSTEM FOR SECRET STANDOFF INSPECTION OF PEOPLE

Andrey Kuznetsov[1] and Alexey Evsenin
Applied Physics Laboratory, Khlopin Radium Institute, Saint-Petersburg, Russia

Abstract. A device for standoff detection of metallic and non-metallic objects concealed under clothing on human body is described. The device, which is based on active interrogation with gigahertz-range electromagnetic waves, can perform secret inspection of moving subjects. It provides both the image of concealed objects and their dielectric characteristics, which can be used to facilitate detection of explosives.

Key words: explosives, standoff, microwaves, gigahertz, dielectric properties

2.1. Introduction

Electromagnetic (EM) waves with frequencies about tens of gigahertz may be primary candidates for standoff detection of explosives concealed under clothing on human body for the following reasons:

- They have high-enough penetrating ability in wet environments (unlike terahertz-range EM waves), enough to penetrate tens of meters of humid air and layers of wet clothing.
- They do not require bulky portals, delicate aiming devices etc, so the potential "suicide bomber" is not aware of the fact that he is inspected.
- All the required components (sources, receivers, electronic components) are cheap and readily available.

Very low power of the EM waves emitted as a discrete set of narrow frequency lines makes the technique completely safe for human health and for electrical appliances. Image resolution of ~1 cm can be achieved, which, though less than that of terahertz or X-ray systems, is enough to recognize the concealed threat, while not presenting privacy problems.

[1] E-mail: APL@ATOM.NW.RU

H. Schubert and A. Rimski-Korsakov (eds.), Stand-off Detection of Suicide Bombers and Mobile Subjects, 5–9.

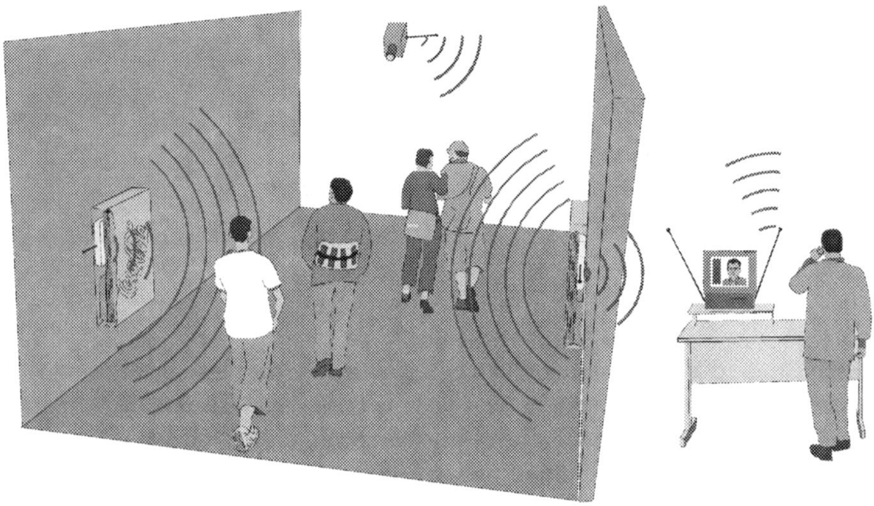

Figure 1. Example of the MS-SRIP installation in a corridor

The proposed Microwave System for Secret Remote Inspection of People (MS-SRIP) is based on active probing with microwaves. The system is intended for real-time detection of metallic and dielectric objects hidden under clothes on body of a moving person (see Figure 1).

2.2. Experimental Results with Prototype

Human body is a perfect reflector of gigahertz-range microwaves, so any object attached to it will be seen as having one or several additional reflecting surfaces. Owing to the fact, that metals (conductors) and dielectrics reflect EM waves differently, the detection system can provide the following information about the inspected target:

- 3D image of objects hidden on human body.
- Classification of detected objects into metals/dielectrics.
- The value of dielectric constant of the object (if it is a dielectric).
- The mass of the concealed object in TNT equivalent.

Dielectric characteristics of several types of real explosives have been determined in dedicated experiments. Though not a unique characteristic of explosives, dielectric constant is an important additional indicator, which allows one to distinguish between concealed explosives and, say, a wallet containing plastic, paper etc.

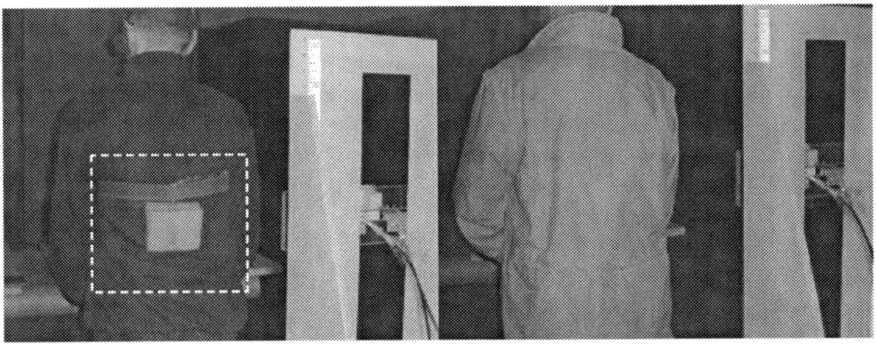

Figure 2. Experimental setup with two antennae. "Suicide bomb" was simulated by wax attached to the body and covered by a raincoat

The laboratory prototype of the inspection system was working in the frequency range 2–8 GHz and emitted coherent microwaves with frequency step dF = 125 MHz. The radiating aperture was synthesized by moving the elementary antenna with uniform step of 2 cm along vertical and horizontal axes within the rectangle with dimensions 60 cm × 40 cm.

The laboratory prototype for detection of explosives concealed on human body is shown in Figure 2. The rectangle on the left photo indicates the boundaries of the inspected region.

A block of wax attached to the body simulated the explosive charge. Dielectric properties of wax are close enough to those of standard explosives.

A 3D image of the wax attached to human body is shown in Figure 3. Since the wax simulator is a dielectric, its image has two surfaces: the first surface is the front border clothing-simulator, the second is the border simulator-body, which is located from the front border at distance, determined by the electrical length of the simulator. Analysis of Figure 3 yields the following conclusions about the found object:

- The object attached to human body is a dielectric.
- Dielectric constant of the object is 2.6 (correct for wax).
- Mass (in TNT equivalent) of the detected object is 1.8 kg.

2.3. Expected Characteristics of MS-SRIP

Expected characteristics of the full-scale system:

- *Frequency range:* several tens of gigahertz.
- *Dimensions of the antenna:* 1 m × 1 m.

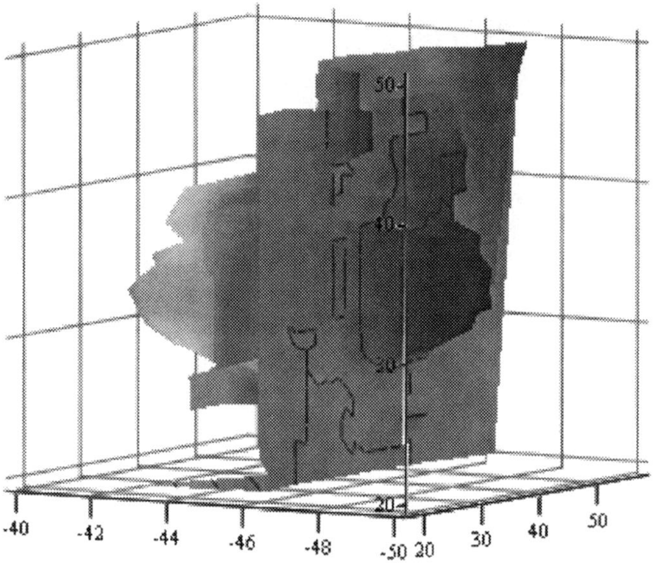

Figure 3. Image of the area of the body containing wax (dashed rectangle in Figure 2)

- *Resolution:* $2 \times 2 \times 1$ cm^3 at 2m; $4 \times 4 \times 1$ cm^3 at 4 m; $10 \times 10 \times 1$ cm^3 at 10 m.
- *Secrecy:* imaging of moving targets; people do not have to stop in front of the device.
- Real-time operation.
- *Safety:* no "real" focusing of microwaves; emitted power less than that of a conventional mobile phone.
- *No privacy issue:* resolution is enough for detection of explosives, but not enough for revealing body details.

2.4. Further Plans and Conclusions

Gigahertz-range electromagnetic waves have been shown to be an adequate tool for standoff detection of "suicide bombs" on human body.

Any useful system for standoff detection of "suicide bombs" must be able to examine moving people in real-time. To achieve that goal, an antenna array has been developed (see Figure 4), which does not contain moving parts and can provide images at a rate of up to a 100 frames per second.

Further work will include:

- further development of methodology of remote inspection of human body by microwave technique; experimental investigation of the maximal detection distance; conceptual design of MS-SRIP system;

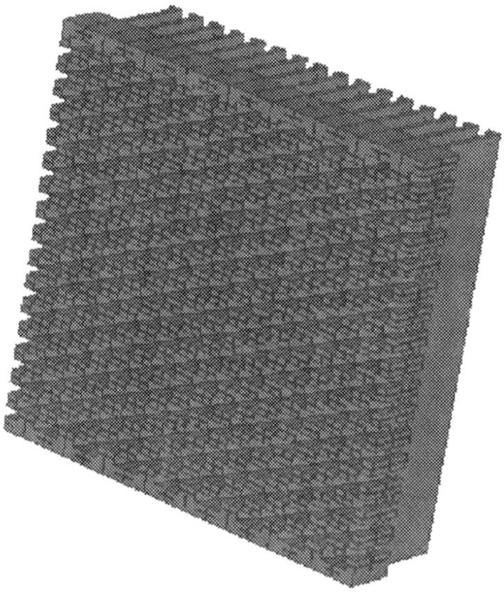

Figure 4. Drawing of the antennae array, which will be a part of the real-time inspection system

- production and testing of a fully functional laboratory prototype;
- production of a full-scale prototype of MS-SRIP; tests and demonstration at test-grounds.

References

Averianov, V.P., I. Yu. Gorshkov, A.V. Kuznetsov, and A.S. Vishnevetski, 2004. Detection of explosives using continuous microwaves, Proc. of the NATO ARW #979920: Detection of Bulk Explosives: Advanced Techniques against Terrorism, St.-Petersburg, Russia, Kluwer Academic Publishers, NATO Science Series, Series II: Mathematics, Physics and Chemistry, Vol. 138, pp. 161–166.

Kuznetsov, A.V. et al. 2003. Portable multi-sensor for detection and identification of explosive substances, Proc. of the International Conference EUDEM2-SCOT: Requirements and Technologies for Demining and the Removal/Neutralization of Unexploded Ordnance, 15–18 September, 2003, Vrije Universiteit Brussel, Bruxelles, Belgium, edited by Sahli, A.M. Bottoms, J. Cornelis, pp. 633–637.

Patent: Russian Federation #2230342 Method of identification of dielectric objects, priority of invention date: 9 September, 2002.

3. SECRET MICROWAVE "DOOR" FOR INSPECTION OF PEOPLE AND LUGGAGE

Igor Gorshkov[1], and Valery Averianov
Applied Physics Science and Technology Center (APSTEC) Ltd, Saint-Petersburg, Russia

Abstract. Devices for secret detection of explosives hidden on human body and in luggage based on gigahertz-range electromagnetic waves are described.

3.1. Introduction

A concept of using active probing with microwave electromagnetic radiation in security-related applications has been developed.

The technology has been implemented in a device for luggage screening.

- Laboratory prototype system for luggage screening has been built.
- Data analysis algorithm has been developed, which allows one to obtain image of the internal structure of the luggage and its dielectric profile, and to detect dielectric as well as metallic objects, thus solving the "thin metallic foil" problem of X-ray and NQR-based systems.
- Experiments with real suitcases and explosive imitators have been conducted.

We propose to use the method tested on luggage for screening human body in order to detect concealed "suicide bombs" and metallic and non-metallic weapons. Use of active probing stepped-frequency continuous gigahertz-range electromagnetic waves and analysis of the reflected and transmitted field would allow one to obtain images of objects concealed on human body and to determine their dielectric characteristics.

The system will be capable of detecting and localizing concealed "suicide bombs", metallic and non-metallic weapons hidden under clothes on human body, in a suitcase, rucksack, or luggage. It can be secretly installed in doorways, as additional sensors in existing portals (e.g. metal detectors), in turnstiles of shops, offices etc.

[1] E-mail: info@apstec.ru

H. Schubert and A. Rimski-Korsakov (eds.), Stand-off Detection of Suicide Bombers and Mobile Subjects, 11–16.

The design of the system for inspection of human body has been developed in accordance with the following guidelines:

- Secret inspection.
- Real-time inspection of moving people.
- Safety for human health and for electric appliances.

3.2. Examination of Luggage

First experiments were carried out with luggage moving on a conveyor belt. The system can be used as an independent inspection tool, inexpensive and real time, or as an additional sensor in conjunction with an X-ray and/or QR systems. Unlike both X-rays and QR, microwave system can visualize very thin metallic foils, which may be used to cover the explosives to make it undetectable by QR method. It can also provide dielectric properties and dimensions of objects inside suitcases, which can be used to facilitate detection with X-rays (e.g. to move from surface densities to volume densities).

Work of the system is based on irradiation of the examined luggage by microwave radiation, detection of transmitted and reflected electromagnetic field, and reconstruction of the 3D image.

The device consists of several (from 2 up to 4) linear antenna arrays located in the form of a portal, through which the conveyor belt passes. The EM field scatter from the luggage moving with constant speed is measured in a wide frequency band during the time sufficient for synthesizing the aperture in horizontal direction. The measured data are digitized and processed to obtain a three-dimensional image.

The above conceptual scheme was implemented in an experimental device working in frequency range 2–8 GHz (Figure 1). In the horizontal direction the aperture was synthesized due to the movement of the examined luggage, and in vertical – by moving the emitting and receiving antennae.

Prior to the experiments, dielectric properties of several types of real explosives were measured in the controlled environment.

The inspected suitcase contained many common objects (Figure 2) like clothing, magazines, photo camera, a pack of wax (imitating explosives), etc. Examples of the obtained images of the suitcase are shown in Figure 3–5.

Figure 3 shows transmission coefficient for the whole suitcase. Figure 4 shows distribution of dielectric thickness across the suitcase (in millimeters), assuming that dielectric constant is $\varepsilon = 3$, which is close the dielectric

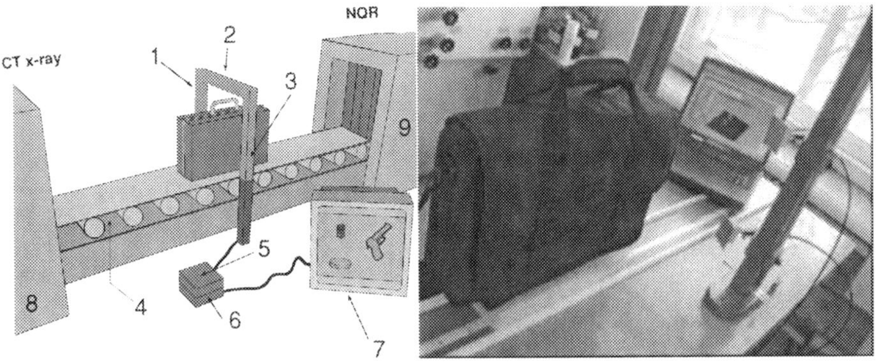

Figure 1. Conceptual design and laboratory prototype of a conveyor-mounted inspection device. 1, 2, 3: antenna arrays; 4: conveyor belt; 5, 6: electronics and data analysis unit; 7: monitor

constant of some explosives. This allows one to estimate the amount of the potentially hazardous material. Figure 5 shows one of the cross-sections of the suitcase, on which metallic construction frame, photo camera and a battery are visible.

According to the experiments, the full-scale system will be able to:

- Construct 3D image of the inside of the suitcase.

Figure 2. The inspected suitcase contained clothing, paper, wax, photo camera, soap, explosives' imitator etc

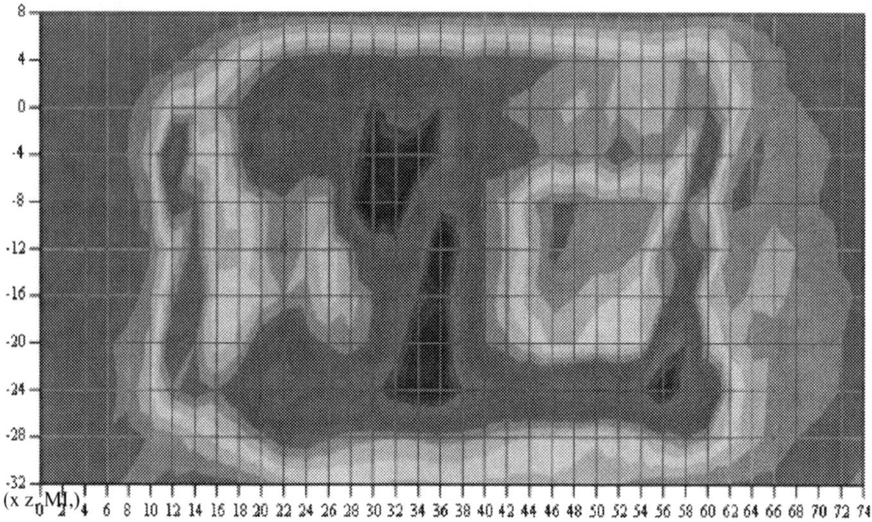

Figure 3. Imaging of metals: transmission coefficient (red: full transmission; blue: no transmission)

- Detect unshielded dielectric inside suitcase and determine its dimensions and dielectric constant, which would facilitate X-ray systems in moving from surface density to real density of the detected object.
- If the explosive is shielded by metallic foil, the system will detect such foil and determine its shape and dimensions.

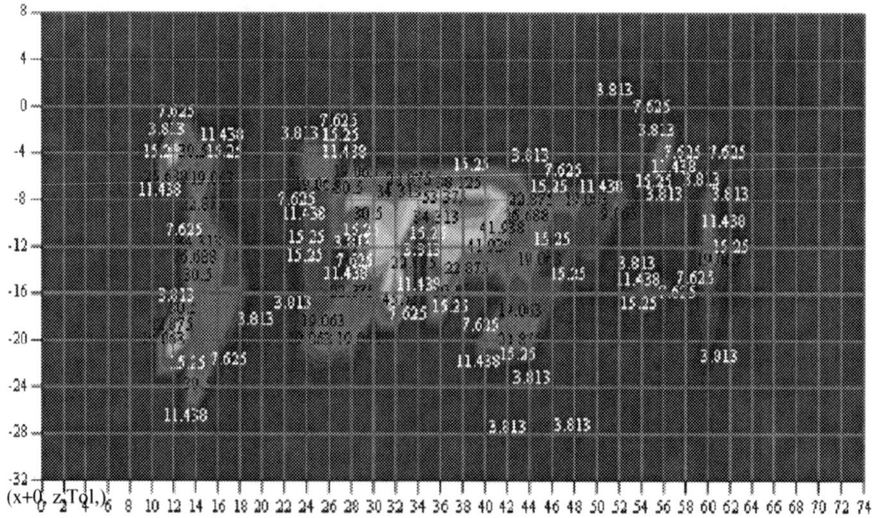

Figure 4. Imaging of dielectrics inside the suitcase (thickness of dielectrics in mm assuming $\varepsilon = 3$, which is close the dielectric constant of some explosives)

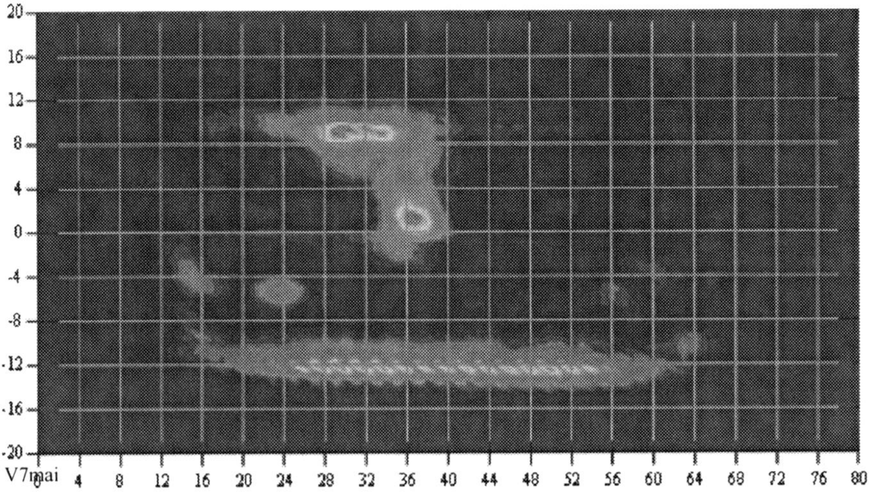

Figure 5. One of the slices of the 3D image of the suitcase

3.3. Detection of Explosives on Human Body

Gigahertz-range microwaves allow one to visualize explosives attached to human body and to determine their dielectric properties and equivalent mass.

The system consists of several antenna arrays secretly mounted in doorways, turnstiles, or in existing portals (see Figure 6). It performs continuous screening in real time of all people passing through the doorway, and the image with added information about dielectric characteristics of concealed objects

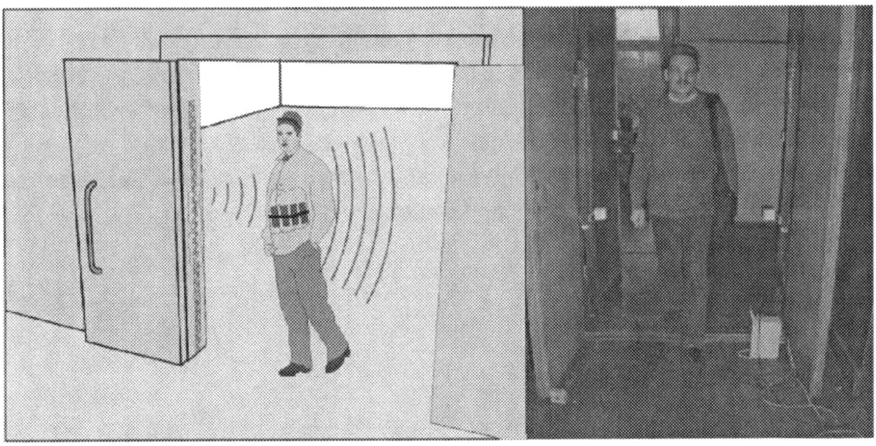

Figure 6. Left: example of mounting options for the Secret Microwave Door. Right: one of the experiments with explosives in carry-on luggage

is passed to the operator. Dielectric properties are used by the automated data analysis procedure to draw operator's attention to the suspicious concealed objects. These properties can be used, for example, to distinguish between explosives and, say, a wallet containing paper money credit cards.

First experiments showed, that the system can detect concealed explosives both on human body, and when they are hidden inside a handbag or a rucksack. Even in these experiments, in which a very simple prototype device was used, detection was done in real time: the person did not have to stop or to slow down (see right part of Figure 6).

Measurements with a laboratory prototype have confirmed the feasibility of using the system for detection of "suicide bombs" on human body and for luggage screening.

Expected characteristics of the full-scale system:

- Two 1–2 m-long transmitter and receiver antennae array at distance 0.5–1 m from the inspected person.
- *Secrecy:* system will be secretly installed in existing construction elements: doorways, turnstiles of shops and metro-stations, offices, etc.
- *Real-time operation:* people do not have to stop in front of the device
- Safety for human health (very low emitted power).

3.4. Conclusions

The system based on gigahertz-range microwaves can perform secret inspection of people with carry-on luggage passing through doorways from distances of about 1 m. It can be secretly installed in doorways, turnstiles, into existing portal devices that use other technologies (e.g. metal detectors).

The principles behind the proposed system have been implemented in a laboratory prototype, which was used to carry out measurements for luggage and body screening. Product engineering of components of system for Secret Microwave "Door" for Inspection of People is under way.

A full-scale prototype working in real time will be then built and tested in different mounting configurations and scenarios.

4. MIP-BASED LOW-COST SENSOR FOR SHORT-RANGE DETECTION OF EXPLOSIVES

G. Bunte[1,3], J. Hürttlen[1], H. Krause[1], H. Pontius[1], M. Schreiter[2], and Jan Weber[2]
[1]*Fraunhofer-ICT, Joseph-von-Fraunhofer-Str. 7, 76327 Pfinztal, Germany*
[2]*SIEMENS AG—CT MM 2, Otto-Hahn-Ring 6, 81730 München, Germany*

Abstract. Fast, selective, sensitive and inexpensive detection techniques are highly required concerning the identification of planned terrorist attacks involving explosive, chemical or biological materials.

State of the art are cost intensive equipments used e.g. in airport portal systems or swapping techniques (Imaging by X-ray, TeraHz-, MMwave, NQR, neutron activation techniques, electronic noses, IMS).

We propose the development of substance specific low cost sensors, which can be installed as self reporting sensors in containers, compartments, wagons etc., used as hand held systems or sniffing devices which analyse the content of explosives in the gas phase.

The basic sensor concept uses mass-sensitive devices coated with molecularly imprinted polymers (MIP). The components are cheap promising low-cost detection devices which could be applied for the passive as well as active sensing of explosives like TNT, DNT or TATP at short-range distances of about one meter. In the case of active sampling the MIP sensor will be combined with a suitable air-sucking system.

Molecular imprinted polymers (MIPs) are highly cross-linked polymers. Like enzymes they possess a high affinity and selectivity for the adsorption of a special target substance (template molecule) which is present only during the MIP-synthesis and is afterwards extracted. This yields to a finished polymer with special and very selective binding sites for the former template. MIPs are very inexpensive, rapid to synthesise and they are already applied as substance specific solid phase extraction materials in the liquid phase.

Preliminary results /1/ of synthesized particulate TNT-specific MIPs showed the effective adsorption capability of the material for TNT vapours. Moreover first experiments using direct UV-initiated polymerisation of TNT-MIPs produced thin films which can be used in combination with different sensor materials.

[3]E-mail: bu@ict.fhg.de

H. Schubert and A. Rimski-Korsakov (eds.), Stand-off Detection of Suicide Bombers and Mobile Subjects, 17–27.

The response time reached with so-called quartz crystal microbalances, QCMs, is in the pg per minute range. Improvement of the MIP capacity will lead to shorter detection times (ca. 1 s).

Overall objective of the proposed study will be to achieve a MIP sensor performance making them usable for short range stand-off detection (1–2 m). The TNT adsorption characteristics of layered MIP-material (pg to ng in seconds-range) has to be improved by varying the synthesis conditions and using QCMs as fast screening tool. The second step of the proposed work will be to coat so-called FBARs, mass-sensitive piezoelectric sensors/arrays (Film Bulk Acoustic Resonators, Cooperation with Siemens) with the improved TNT-specific MIP. Realistic measurements will demonstrate the performance of the cheap sensor prototype.

4.1. Introduction

Recent terrorist attacks like the bombing in New York at 11.9.2001, suicide bombings in the middle east or e.g. in London Underground transportation systems in July 2005 show that the today used detection techniques for explosives are inadequate and have to be improved as well as new counter-action/detection concepts have to be developed.

State of the art are cost intensive equipments used e.g. in airport portal systems or swapping technologies (Imaging by X-ray, TeraHz-, MMwave, NQR, neutron activation techniques, electronic noses, IMS). Under test status are also so-called handheld electronic noses. Disadvantage of the IMS-technique is the necessity of high investment costs for the instruments and the use of vapour detection or thermal desorption of sampled particulates by special swaps. X-ray techniques only detect metal-containments instead of the explosive substance itself and furthermore X-rays cannot detect biological and chemical agents. Concerning the neutron activation techniques as well as for NQR very high investment costs for the instrumentation are compulsory while today only one or two portable systems are commercially available. Moreover neutron activation and NQR techniques have typically very high detection limits of several grams (e.g. 50–100 g for TNT). Explosive detectors based on so-called electronic noses are expensive, big-sized, heavy and need a high air-volume for trapping and enrichment of the analyte in order to realize their limits of detections (ppb-range). Current electronic noses are unable to truly identify an explosive substance but use suitable pattern recognition algorithms for the detection and need a person to be driven.

Direct counter-action of suicide bombers will require the development of the fast, sensitive and long-range specific detection techniques (>10–100 m) in order to have a realistic chance to stop the suspect person before a bombing could occur. Another way to proceed to counteract possible terrorist attacks would be to develop suitable self-reporting sensor networks able to detect explosives plumes of suspect persons or transported goods in a very sensitive and rapid way. Like nowadays everywhere installed fire/smoke sensors these sensors should be able to give an information about the position and/or movement of suspected subjects or objects. Most challenging prerequisite for such self-reporting sensors/networks will be the necessity of suitable micro sensors that can detect targeted explosives components very rapid, sensitive and reliable preferably without the need for costly pattern recognition procedures/software modules as well as trapping devices for sample enrichment.

Instead of today used technologies mass-sensitive devices coated with molecularly imprinted polymers (MIP) seem to provide promising low-cost detection devices for use in selfreporting sensor networks for the surveillance of public areas or as on-line/in-line sensors in different vehicle transportation systems (cars, trucks, containers etc.) in order to directly detect and warn for possible terrorist threats by hazardous components. Overall objective of the proposed study will be to achieve a MIP sensor performance making them usable for short-range stand-off detection (1–2 m). Synthesis and coating conditions will be modified to improve selectivity and sensitivity to achieve the sensor concept. The TNT-adsorption and desorption characteristics of the MIP must be analysed to evaluate the detection limits and selectivity. So-called FBARs, mass-sensitive piezoelectric sensors/arrays (Film Bulk Acoustic Resonators) will be coated with the improved TNT specific MIP and their TNT adsorption performance will be tested. Suitable coating techniques should be identified to transfer the basic sensor concept to a potential manufacturing process. The following chapters will describe the current results of our MIP research as well as the proposed project working steps.

4.2. The Technique of Molecular Imprinting

The technique of molecular imprinting allows the formation of specific recognition sites in macromolecules. In this process functional and cross-linking monomers are copolymerised in the presence of a target analyte (template). The functional monomers form a complex with the imprint molecule and in the following polymerisation the functional groups are held in position by the highly crosslinked structure. Subsequent removal of the template reveals binding sites that are complementary in size and shape to the analyte. The complex between monomers and template can be formed via reversible covalent bonds or via non-covalent interactions like hydrogen bonds (see Figure 1).

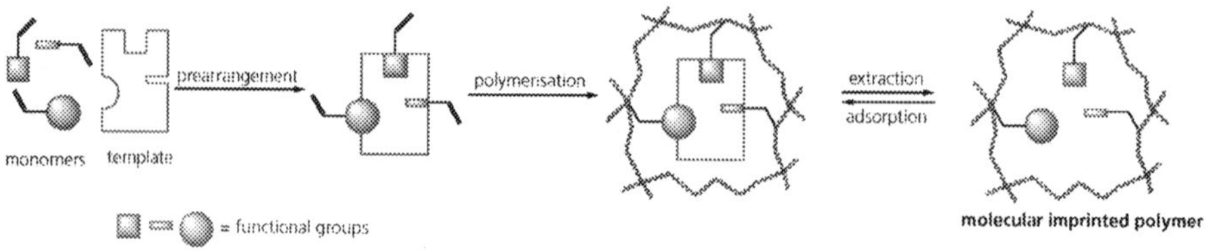

Figure 1. Principle of molecular imprinting

4.3. Synthesis of Porous Beads

For TNT as template it is not possible to use the covalent approach. Furthermore with the non-covalent approach we are able to use a large pool of functional monomers that are commonly used in the field of molecular imprinting. Because of the known problem that nitroaromatics are weak hydrogen bond acceptors [1], we tested several acrylates with different functional groups as monomers to find the one with the best interaction. Ethyleneglycoldimethacrylate (EGDMA) was used as crosslinker (see Figure 2).

It has been reported that the crushing and sieving steps used after traditional bulk polymerisation can break the imprinted sites in the obtained polymer [2]. This process is also labor intensive and wasteful. In order to avoid these problems, the polymers used in the present work were prepared by suspension polymerisation [3]. In a typical experiment 2 mmol of the template TNT, 8 mmol of the functional monomer and 50 mmol EGDMA were dissolved in 25 ml of chloroform. Then the initiator AIBN (200 mg) was added to this mixture and sonicated to dissolve. A low molecular weight polyvinyl alcohol was used as emulator and dissolved in 120 ml of water by stirring at 60°C under nitrogen atmosphere with a mechanical stirrer. After cooling to room temperature, the organic mixture was admitted to the flask at 600 rpm under a gentle stream of nitrogen. Then the temperature was raised to 60 °C and the polymerisation was allowed to proceed for 24 h. The microspheres were washed with water, chloroform, toluene and acetone and dried in vacuum. Non-imprinted polymers were prepared in the same way without the addition of TNT.

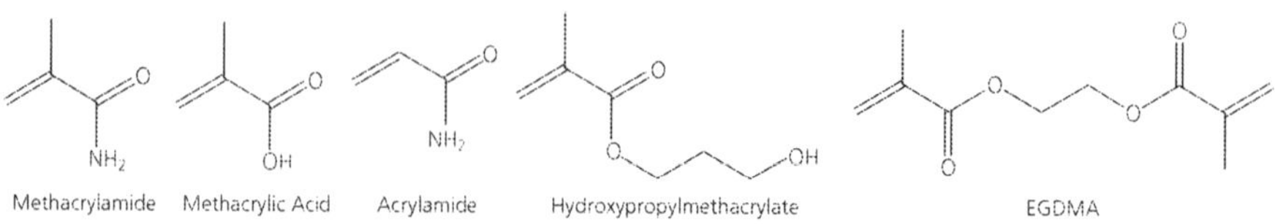

Figure 2. Used functional monomers and crosslinker

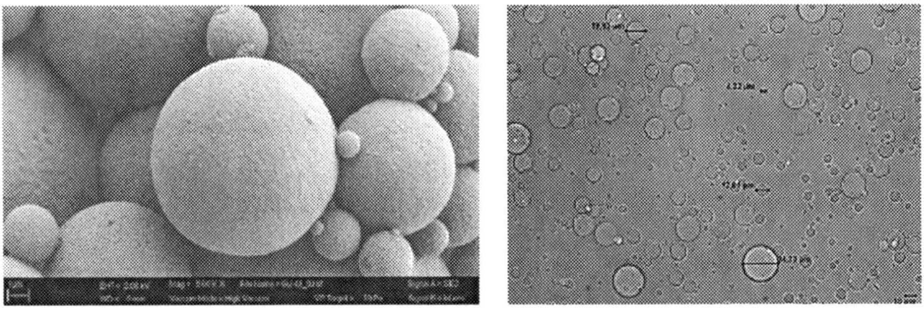

Figure 3. SEM and light microscopy of polymer beads

The synthesized porous beads were characterized by scanning electron microscopy and BET adsorption tests using nitrogen physisorption for the determination of the specific surface area of the MIPs. Like shown in Figure 3 the synthesized MIPs have a diameter of about 10–20 μm. The specific surface area is typically about 200–400 m^2/g.

4.4. Removal of TNT and Performance of Porous Beads

In order to test the performance of the TNT imprinted polymers the template molecules have to be removed. Because of the application of the MIPs as coating for a TNT vapour sensor the template should be removed as much as possible. Furthermore nonreacted synthesis agents should be removed from the MIPs. After washing the imprinted polymers with different solvents (water, methanol, toluene, acetone or chloroform) to purify the MIPs from non polymerised monomer and cross-linking agent different methods were tested to remove the template from the polymer:

- Soxhlet extraction or treatment in an ultrasonic bath using chloroform and subsequent drying under vacuum in an exsiccator.
- Following soxhlet or ultrasonic extraction with supercritical carbon dioxide was used at varying pressures and temperatures or alternatively the MIPs were given into a drying oven at different temperatures (80 °C, 100 °C, 120 °C, 150 °C or 200 °C).

Gas chromatography with mass selective detection (GC-MSD) in combination with indirect solid phase micro extraction (SPME) was used for quantifying the remained template amount, purification and the performance of the MIPs. Following analysis equipment was used: HP 5890 GC and HP 5971A MSD, HP-5MS column (length 30 m, ID 0.25 mm, film thickness 1.0 μm. As carrier gas helium 6.0 was used. In order to determine potential

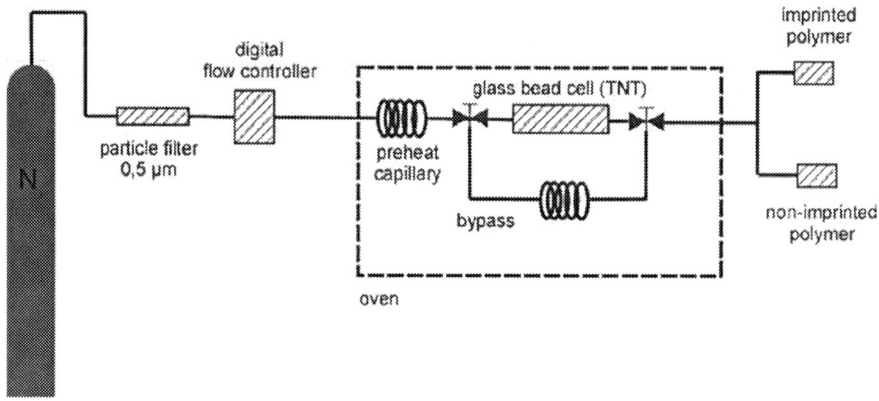

Figure 4. Schematic build-up of the TNT-vapor generation plant

changes in the morphology of the polymers due to the different applied removal procedures the specific surface area and the pore volume were again measured by N_2-adsorption.

Results showed that the best method for purification of the porous beads was soxhlet extraction followed by treatment with supercritical CO2 at mild conditions (150 bar, 50 °C). At least a removal of >99.7% of the template was achieved. Heat treatment at 200 °C lead to the assumingly best/highest TNT-removal but BET measurements showed that the surface area of the beads decreased from 200 m^2/g to 5 m^2/g due to the starting decomposition of the polymer.

After purification, the MIPs have been loaded with TNT vapour to prove the function of the MIPs. For this reason a TNT-vapour generator (Figure 4) was developed. Typical TNT vapour concentration produced was about 4 µg/l (350 ppb). For the performance tests usually several grams of MIP were used. Uploading tests were always made for TNT-imprinted MIPs in comparison to the non-imprinted polymer.

Figure 5 shows the results of the GC-MSD measurements after treatment of imprinted and non-imprinted polymers with TNT vapour for polymethacrylic acid (PMAA) and polyacrylamide (PAA). Independent of the used monomer the unimprinted MIPs only absorb very low TNT amounts. As desired the TNT-imprinted MIPs possess a better ability to bind TNT from the gas phase whereas PMAA shows a more pronounced sensitivity for TNT as PAA.

Preliminary measurements concerning the potential cross-sensitivity of TNT-imprinted MIPs using e.g. 2,4-DNT as test gas provided no measurable absorption amounts as well as DNT-imprinted MIPs did not show measurable affinity for TNT.

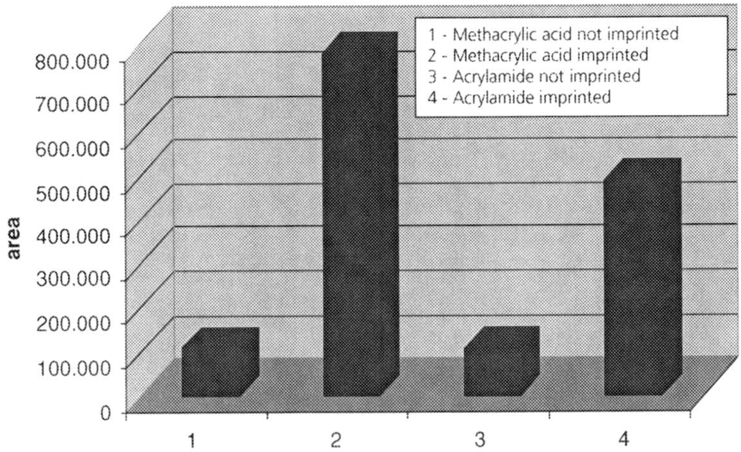

Figure 5. Comparison of the GC-MSD peak areas per mg MIP of particulate TNT-imprinted and unimprinted MIPs (1 and 2 PMAA, 3 and 4 PAA) after TNT vapour treatment

4.5. Preliminary Results for MIP Coatings from Thin Film Polymerisation

For an application of MIPs as specific sensor coatings they should preferably be polymerised as thin films on the sensor surface. In this regard spay coating experiments were performed in combination with fast UV-activated polymerisation of thin films directly on so-called QCMs, quartz crystal microbalances. First experiments using direct UV-initiated polymerisation of TNT-MIPs produced thin films which can be used in combination with different sensor materials. Performance tests of the produced TNT-imprinted and unimprinted MIP coatings were realised using a commercial gaslab (ifak) in which up to eight QCMs can be measured simultaneously under constant temperature (45 °C). Used quartzes have a diameter of nearly 8 mm and a ground frequency of 10 MHz. Mass increases on the QCM will yield to a decrease of the frequency and vice versa.

Exemplarily Figure 6 shows some first test results for TNT imprinted and unimprinted PMAAM coatings. Like detected for the particulate MIPs only the TNT-imprinted MIP coatings show the desired TNT-specific absorption tendency. The measured TNT uptake was about 1.2–2.5 pg per hour.

Due to modification of the synthesis conditions for other coatings the TNT absorption was increased by a factor of 40 (Figure 7). Preliminary results show that TNT specific MIPs could as well be synthesised as thin films. The response time reached with the quartz crystal microbalances, QCMs, is in the pg per minute range. Further improvement of the MIP capacity will lead to shorter detection times (ca. 1 s).

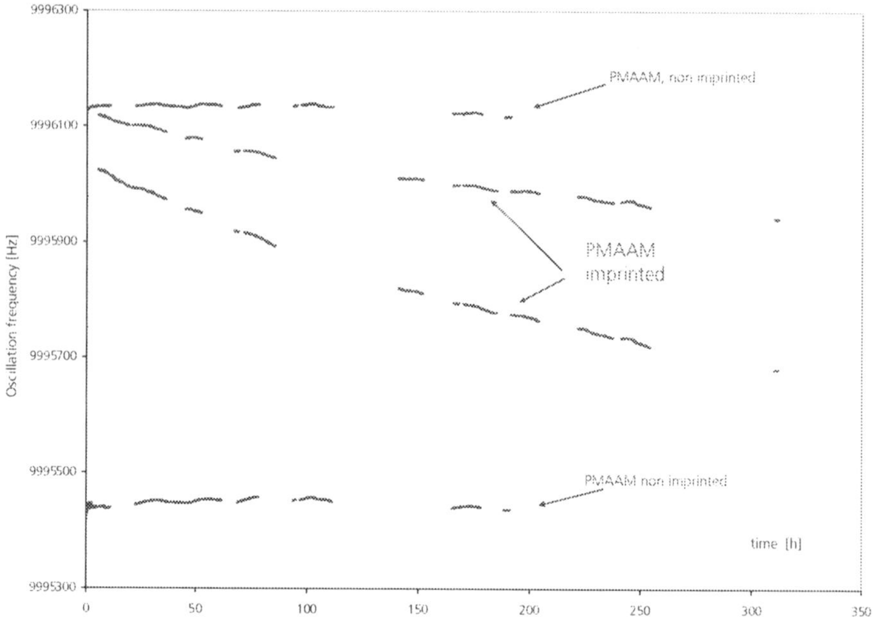

Figure 6. Frequency measurements of four QCMs coated with TNT imprinted and unimprinted PMAAM during TNT vapour treatment

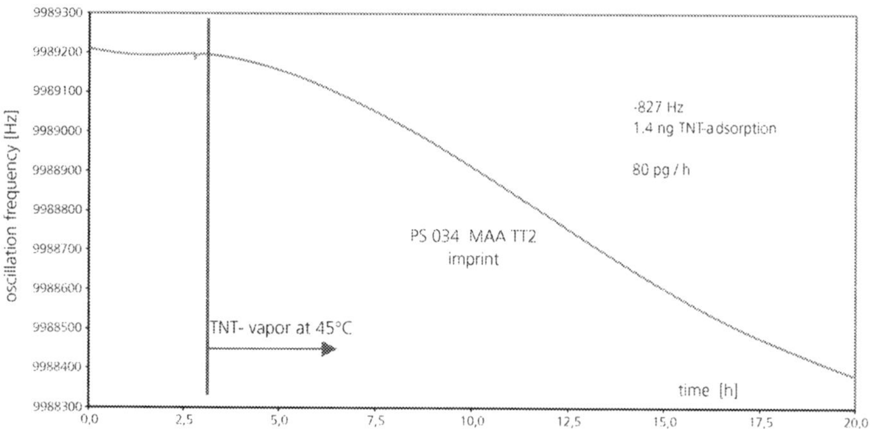

Figure 7. Frequency decrease of a QCM with improved MIP during TNT vapour treatment

4.6. Properties and Potential Use of Film Bulk Acoustic Resonators (FBAR)

Piezoelectric devices convert acoustic into electric energy and vice versa. They show an excellent frequency stability and have therefore been applied as tools for high precision determination of frequency and time in clocks or electrical transmitters for a long time already. During the fifties Sauerbrey [5] demonstrated the direct correlation of frequency change and a thin film mass attachment and so their use extended into mass sensing applications. Since then their importance in this area has steadily been increasing making them today standard devices which are widely used in Bioanalytics and gas sensing. Nevertheless the principle composition of such piezoelectric microbalances has remained unchanged and consists of a quartz disk that is obtained by cutting from the whole crystal and that is clamped thus forming a membrane.

During the nineties thanks to a rapid development of semiconductor technology, fabrication of very thin piezoelectric films of high quality was achieved. With such films piezoelectric filters working at frequencies of several GHz were obtained and are now employed in mobile communication devices. Within the EU-Project "PISARRO" [6] with the participation of Siemens and others, FBARs (Film Bulk Acoustic Resonators) were developed for application in the area of gas- and biosensing and an excellent performance was obtained [7, 8].

FBARs have significant advantages over quartz crystal microbalances. Due to fabrication with standard semiconductor technologies they can be made very small (down to a few micrometers) and thus manufacturing costs are very low. Fabrication is compatible with standard frontend CMOS processes, so sensor and readout circuitry can in principle be processed on the same wafer. Finally FBARs have an excellent suitability for array integration. By providing the individual pixels with different functional coatings the device can be made sensitive to a number of gases or substances. High selectivity can be achieved even if the specificity of the individual functionalisations is rather week by pattern recognition of the output signals of a large number of pixels.

The functional principle of the FBAR as it is developed at Siemens is shown in Figure 8. It comprises three elements. The first is the piezoelectric transducer which consists of the piezoelectric layer and bottom and top electrode. The second is an acoustic cavity which ensures the acoustic energy to stay in the top layers thus increasing the sensitivity. This is achieved by an acoustic Bragg mirror. Finally the functional coating provides the selectivity of the device. When a molecule is attached to the functional coating the mass of the piezoelectric device increases and the resonance frequency decreases. This frequency change can be read out.

The performance of the sensor is not only ruled by selectivity as mentioned above, also the sensitivity plays an important role. Sensitivity is given by the change of the measurand (the frequency) per mass attachment per unit area. Together with the frequency fluctuation a value for the minimum detectable

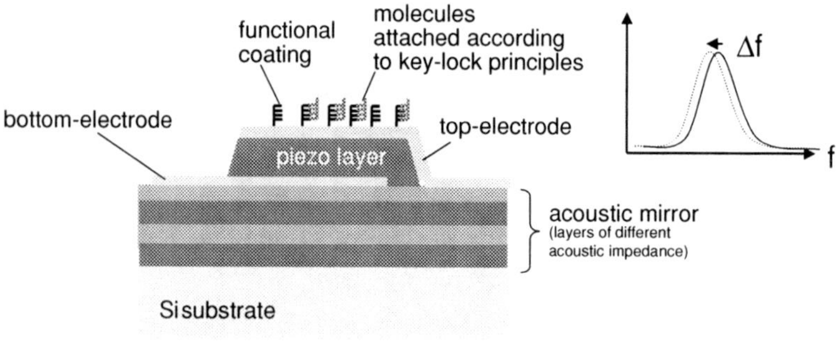

Figure 8. Schematic view of the FBAR device comprising a piezoelectric layer plus electrodes, an acoustic mirror for confinement of the acoustic energy and a functional coating for device selectivity. Due to mass attachment a negative shift of resonance frequency is obtained

mass attachment can be obtained. The frequency fluctuation was obtained from stability measurements and is 500 Hz for a 2GHz-FBAR. The sensitivity for this device is 2 kHz · cm^2/ng. With these values, taking the threefold noise as minimum detectable frequency shift, the minimum mass attachment is 0.75 ng/cm^2.

4.7. Proposed Project

We propose the development of substance specific low cost sensors, which can be installed as self reporting sensors in containers, compartments, wagons etc., used as hand held systems or sniffing devices which analyse the content of explosives in the gas phase.

The basic sensor concept uses mass-sensitive devices coated with molecularly imprinted polymers (MIP). The components are cheap promising low-cost detection devices which could be applied for the passive as well as active sensing of explosives like TNT, DNT or TATP at short-range distances of about one meter. In the case of active sampling the MIPsensor will be combined with a suitable air-sucking system.

Overall objective of the proposed study will be to achieve a MIP sensor performance making them usable for short range stand off detection (1–2 m). The TNT adsorption characteristics of layered MIP-material (pg to ng in seconds-range) has to be improved by varying the synthesis conditions and using QCMs as fast screening tool. The second step of the proposed work will be to coat so-called FBARs, mass-sensitive piezoelectric sensors/arrays (Cooperation with Siemens) with the improved TNT-specific MIP. Realistic measurements will demonstrate the performance of the cheap sensor prototype.

Acknowledgement

Financiation of the presented MIP-concerning work by the German MoD is kindly acknowledged.

References

[1] Baitinger, W.F., P.R. Schleyer, T. Murty, and L. Robinson, 1964. Tetrahedron, 20, 1635–1637.

[2] Mayes, A.G. and K. Mosbach, 1996. Anal. Chem., 68, 3769–3774.

[3] Lai, J.-P., X.-Y. Lu, C.-Y. Lu, H.-F. Ju, and X.-W. He, 2001. Anal. Chim. Acta, 442, 105–111.

[4] Bunte, G., J. Hürttlen, G. Unkelbach, K. Niedergall, and H. Krause, Imprinted polymers for gas phase detection of TNT, MIP 2004—Symposium, 12–15 September 2004, Cardiff, UK.

[5] Sauerbrey, Verwendung von Schwingquarzen zur Wägung dünner Schichten und zur Mikrowägung, 1959. Z. Phys. 155, 206–222.

[6] Piezoelectric Sensing Arrays for Biomolecular Interactions and Gas Monitoring (PISARRO), with Siemens AG, Technobiochip, E+E Elektronik, VTT Automation, Nat. Inst. Lasers, Cranfield University and University of Linz (JKU).

[7] Gabl et al., 2004. First results on label-free detection of DNA and protein molecules using novel integrated sensor technology based on gravimetric sensor principles, Biosensors and Bioelectron., 18, 615–620.

[8] Weber, J., W.M. Albers, J. Tuppurainen, M. Link, R. Gabl, W. Wersing, and M. Schreiter, 2006. Shear mode FBARs as highly sensitive liquid biosensors, Sensors Actuators A, 128(1), 84–88.

5. SCANNED MULTIPLE (ELECTRO-)CHEMICAL SENSORS LOCATED AT CRITICAL INFRASTRUCTURES

G. Sakovich[1], O. Tolbanov[1], A. Vorozhtsov[1], M. Krausa[2], and N. Eisenreich[2]
[1] *The Institute of Problems of Chemical and Energetic Technologies, Russia*
[2] *Fraunhofer Institut für Chemische Technologie (ICT), Pfinztal, Germany*

5.1. Scenario

A new threat by terrorism has reached World. Initiated explosives have already killed thousands of people and injured several tens of thousands. Infrastructural facilities like airport, railway stations, city halls, water supply, etc. could be preferred targets involving up to thousands of people. Assuming, methods will be found to early detect explosives with help the sensors placed at key positions, an immediate response will prevent far ranging consequences.

Terroristic threat requires standoff detection:

- We have to protect our inhabitants;
- We have to protect our infrastructures;
- We have to develop new technologies.

Problems to be solved:

- Be present everywhere;
- Early detection and identification;
- Sensitivity, reliability and cost effectiveness.

A threat by terrorists operating as single persons or groups in cities can be only countered if surveillance by cheap anonymous sensors works at place they must pass on their way to their operation. Carrying the explosive charge in substantial masses will leave trace of extremely small vapour of the material. Cheap sensors of extremely high sensitivity will detect this trace if they are located at key places they have to pass.

To summarize the approach:

5.1.1. Detect the Situation—Identify the Threat

- Install multiple detector at critical infrastructure they might pass;
- Continuously scan and evaluate the sensor status;
- Sensitivity, reliability and cost effectiveness.

H. Schubert and A. Rimski-Korsakov (eds.), Stand-off Detection of Suicide Bombers and Mobile Subjects, 29–34.

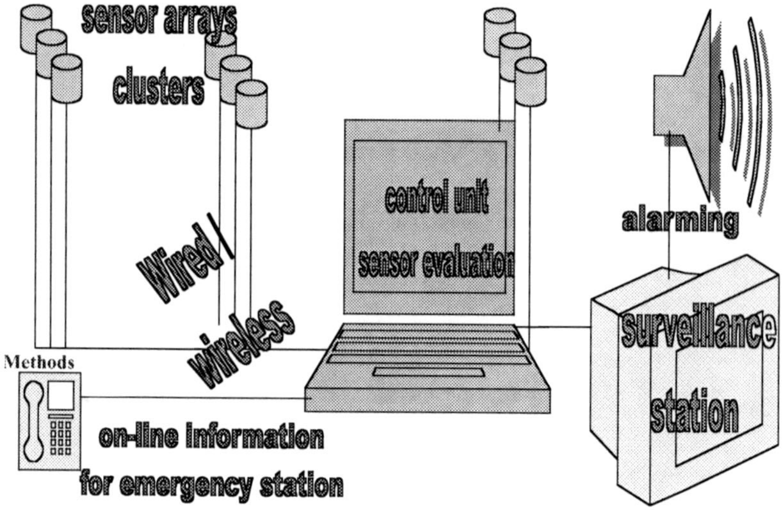

Figure 1. Scheme of the multi-sensor approach

5.2. Objectives

It is the objective to develop a concept for a scanned multiple (Electro-) chemical sensor systems for an early detection, warning and recording system for propagating single persons or groups carrying explosive for terroristic operations (see Figure 1). It maintains surveillance also in case of a disaster and initiates emergency actions. It combines

- multi-sensor techniques o various techniques;
- Sensor networking including on-line control, tracing the paths of terrorists when passing infrastructures of bus, underground, railway stations, airports, sports arenas, central places etc;
- advanced statistical data pretreatment and evaluation taking advantage from multi-sensors;
- increasing the reliability by redundant multi-sensor evaluation and immediate alarm/response.

To achieve sensitivities in total to 1–10 ppt at 1–3 m distance of the vapour source.

5.3. Approach for Realization of the Specifications

The status of single sensors is actually:

- *Sensitivity:* 10–100 ppb,
- *Selectivity:* depends on problem and type

⇨Reliability of operation depends on both aspects.

Problem to be solved by new developments => <1–10 ppt:

Therefore it should be strongly enhanced sensitivity, selectivity and reliability!

An approach to use the multiple sensor combination and advance methods of and statistical data evaluation and data fusion might solve the problem.

Two types of sensors—electrochemical and metal oxide semiconductor based ones—will be substantially improved in their sensitivity and selectivity by

- Nano-structured materials of their electrodes—improvement about a factor 100—1000;
- applying arrays or clusters of sensors to enable an effective statistical data evaluation—ca 100;
- sampling by sucking recoverable microstructured capillary adsorbers — ca 100–1000.

A network will link these sensor clusters and perform special scan cycles to increase the reliability of the sensor information and to decide on stages of alarm and recording the situation.

The development will initiate a sensor head development of 2 types of sensor are to be developed to increase sensitivity in arrays and cluster configuration:

- a MOS-based sensors—porous nanosystems, containing nanoparticles;
- Electrochemical sensor;
- Sampling system based on adsorption/desorption techniques.

Substances to be detected are TATP for start and TNT, DNT, HMX, RDX

The MOS-based sensor operate on the following principles (see also Figure 2):

5.4. Chemical Sensors: Metal Oxide Sensors

5.4.1. PRINCIPLE

Adsorption/desorption processes at the surface, conductivity changes of the following substances

SnO_2, ZnO, WO_3, TiO_2 etc.

Impact on the Sensitivity and selectivity is given by material, particle size, layer thickness, additives and temperature.

Selective sensor properties determines by

- Material, width and morphology of film (see Figure 3);
- Range of temperatures and exposure time of thermocycling regime;
- Activity and selectivity of catalytic membrane.

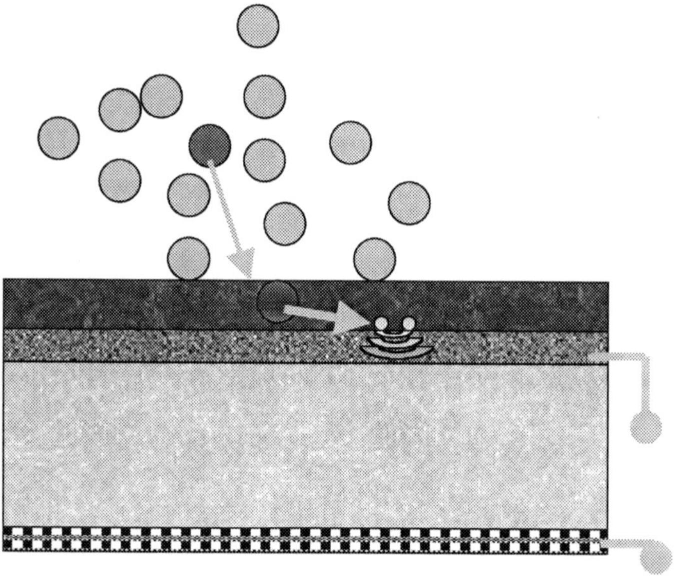

Figure 2. Physical sensor principle

The following results are expected:

- to optimize an initial location and number of sensors for specific objects with the purpose of reliable identification adequately any possible position of source of explosive vapors on known distance: (1.5–3 m) sensitivity of gauges;

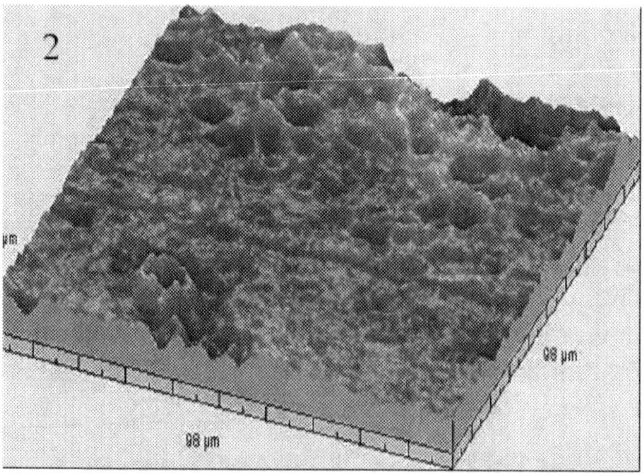

Figure 3. Morphology of sol–gel films—cells of gas sensors

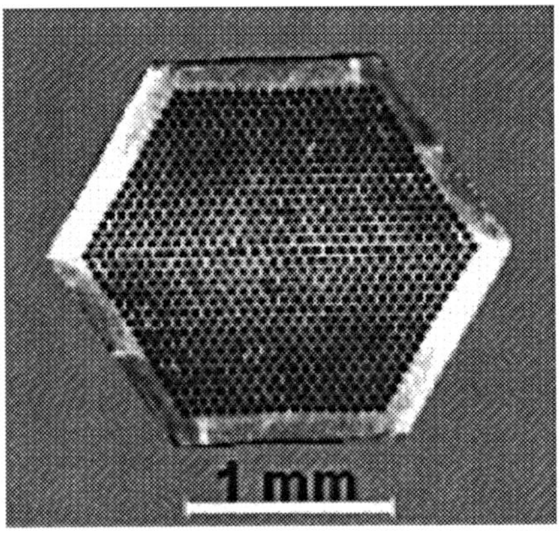

Figure 4. Example of a sampling filter which meet most of the requirements

- with a high degree of probability to identify a position of concrete source in conditions of limited time (the order 3–7 s) taking into account non-stationary disturbances of air streams (passing trains of the underground stations, system of ventilation of premise).

5.4.2. SAMPLING

Demands on sampling methods for stand off detection will increase the concentration to be detected by the sensor: fast, small and repeatable.

The methods of resolution are:

- microstructured set-up;
- capillary tubes (large surfaces areas, high porousness);
- materials: polymers, metals and glasses;
- fast heating/cooling cycles;
- investigations on enhancement of selectivity by sampling (see Figure 4).

5.5. Sensor Combination and Data Evaluation

5.5.1. SENSOR COMBINATION

Simulation of smell sense by combination of MOS and electrochemical sensors will achieve approx. 30 ppt for TNT (actual). Better discrimination will be possible by than by an electrochemical sensor There were already results available which were by sensor combination to 30 ppt obtained for TNT.

Measurement technology development has to include (see scheme in Figure 5):

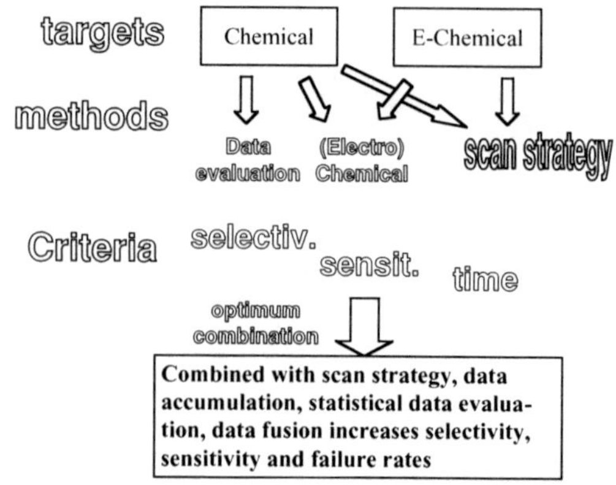

Figure 5. Sensor combination and data evaluation

- *Electronic configuration:* amplification, smoothing, data pretreatment;
- *Control unit and statistical data evaluation and calibration* with sample sets scan strategies, data fusion: signal integration, accumulation, signal evaluation/correlation;

Application of Multi-component analysis will be applied as it is standard for NIRS. Methods are: PCA; PLS, MLR etc.

Calibration by sample sets have to be produced as main input o the data evaluation. It is important also to include sample sets of substance not to trigger but might have a similar signature like the explosive to be detected.

Prototype development and testing will be steps to realize the technology:

- Establishing of specifications related to a scenario;
- Improvement of the single sensors and arrays;
- Development of the measurement electronics;
- Realisation of the sensor combination;
- Apply the statistical data evaluation codes;
- Define and prepare the calibration sample sets;
- Implement a control unit with data fusion software and scan strategies;
- a prototype with basic components for verification;
- verification by tests.

The result will be a prototype of hardware and on-line software package.

6. REAL-TIME DETECTION OF IED EXPLOSIVES WITH LASER IONIZATION MASS SPECTROMETRY

Henric Östmark[1,3] Sara Wallin[1], Anna Pettersson[1], and Harald Oser[2]
[1]*Swedish Defence Research Agency, FOI, Grindsjön Research Centre, SE-147 25 Tumba, Sweden*
[2]*SRI International, 333 Ravenswood Avenue, Menlo Park, CA 94025-3493, USA*

Abstract. The ESSEX system (Extremely Sensitive and Selective Explosives detector) is a system that can be used for standoff or remote detection of suicide bombers at a distance of up to a few meters via the trace amounts of explosives present in vapor phase around explosives with a high vapor pressure. The ESSEX detection method is based on laser ionization mass spectrometry (LI–MS). Direct sampling of air with explosives in vapor phase will allow detection of explosives with relatively high vapor pressure, such as TATP (suspected to have been used in the recent suicide bombings in London as well as numerous suicide bombings in Israel) and EGDN (normally found in dynamite explosives such as the ones used in the Madrid bombings). The LI–MS method has a potential to detect all or nearly all explosives. For explosives with low vapor pressure (e.g. RDX, HMX, PETN, AN) it is probably necessary to sample particles as well. Other aspects of the future potential for this method are that the number of explosives that the method targets can be increased as needed. The method is very suited for data fusion and processing, making it possible to incorporate artificial intelligence (AI) into the system. The uniqueness of LI–MS methods lies in its low detection limit, possibility of real-time detection, its versatility regarding the number of possible substances to detect and its unique selectivity, leading to very few false alarms. The ESSEX concept is appropriate for indoor and outdoor applications, it is benign to humans and property and can be applied in an unnoticeable way. Laser ionization mass spectrometry and specifically resonance enhanced multi photon ionization (REMPI) is an ultra-sensitive, highly selective analytical technique that can identify and quantify vapor-phase constituents at parts-per-trillion (ppt) levels. The REMPI technique combines the principles of optical spectroscopy and mass spectrometry to provide a two-dimensional detection scheme that yields a high degree of chemical sensitivity

[3]E-mail: henric.oestmark@foi.se

H. Schubert and A. Rimski-Korsakov (eds.), Stand-off Detection of Suicide Bombers and Mobile Subjects, 35–50.

and selectivity. This method has proven effective for the detection of one trace compound at ppt levels in the midst of others. The sensitivity and specificity for REMPI can be dramatically improved by using cooling with a supersonic jet based inlet. This cooling greatly simplifies the spectra, producing narrower and stronger electronic transitions and allows sampling directly from air. This paper describes the principles of LI–MS for standoff detection the trace amounts of explosives present in vapor phase around explosives with a high vapor pressure as well as the future potential of the method.

Key words: resonance enhanced multi photon ionization, jet-REMPI, explosives detection, energetic materials, improvised explosive devices, IED, laser ionization mass spectrometry, LI–MS

6.1. Introduction

The terrorist threat to society in the form of attacks on airports, trains, and other public arenas is an increasing problem. Not only have terrorist attacks been occurring more frequently, they also introduce new threats in the form of new explosives, explosives-related materials, and improvised explosive devices (IEDs). This escalation calls for an explosives detection method that is fast, precise, and easily expandable to incorporate new threat substances as they emerge. The problem with suicide bombers is also an increasing threat, not only in countries like Israel where this has been a problem for a long time but also in other parts of the world, e.g. London. The risk of suicide bombings also poses an additional requirement for the detection to be unnoticeable to persons under inspection.

ESSEX will employ the laser ionization mass spectrometry (LI–MS) technique, which was developed on the basis of resonance enhanced multi photon ionization (REMPI); a fast, ultra-sensitive and highly selective ionization technique that is able to identify and quantify vapor-phase constituents at parts-per-trillion (ppt) levels [1]. ESSEX can be used in both indoor and outdoor applications and is benign to humans and property in the interrogated area. Like other trace detection methods, it is appropriate for screening of people and inspection of items carried onto airplanes or trains or into buildings or public arenas. When used for high vapor pressure explosives it can easily be made unnoticeable to persons under inspection. There are no privacy issues associated with this method since it is non-intrusive.

Current means of trace detection such as ion mobility spectrometry (IMS) (as discussed in a report by the National Research Council [2]), gas chromatography/electron capture detector (GC/ECD) and gas chromatography–mass spectrometry (GC–MS), are limited in several ways. With IMS, the number of interrogation targets is very limited and the false positive rate is relatively high. The GC–MS method is specific but much too slow for screening purposes, and the GC/ECD method needs even longer interrogation times to be selective.

The handling of explosives leaves trace residues on hands and clothes, even when meticulous care is exercised to avoid contamination. Trace residues will also be present on secondary objects such as suitcases, boarding passes, and so on. Therefore, detection will in most cases be possible even when the explosive is concealed. Trace amounts of explosives are present in gas phase, adsorbed on particles, or in particle form. Trace detection systems such as the one described in this paper are therefore capable not only of detecting the presence of bulk quantities of explosives but also alerting screening personnel to security concerns about a passenger who has had contact with explosives.

Sampling of air with a direct inlet to detect explosives present in vapor phase will allow detection of explosives with relatively high vapor pressure, such as TATP (suspected to have been used in the recent suicide bombings in London as well as numerous suicide bombings in Israel) and NG or EGDN (normally found in dynamite explosives such as the ones used in the Madrid bombings). In the case of explosives with very high vapor pressure it is possible that the particles will evaporate off before the detection event. However, in this case it is likely that large amounts of explosives vapor will be present around the explosive charge since it is very difficult to make air-tight seals on a container. For detection of explosives with low vapor pressure a particle sampling system needs to be developed.

ESSEX will directly sample the air, making the sampling process very fast. Soaked-up vapor from the person or item under inspection will be analyzed by means of highly selective, molecule-specific laser ionization followed by mass measurement. The resulting two-dimensional data will be compared to a database and the result will be processed automatically into a "yes/no" decision on whether an explosive is present so that no human interpretation is required. The high specificity of the data admits advanced data processing, making it possible to perform situation and threat analysis even under circumstances of uncertain, incomplete, and contradicting information.

ESSEX will combine high sensitivity (limit of detection, or LOD, in the low-ppt range) with fast sample processing that allows detection within a few seconds. Furthermore, the system can be upgraded to detect more substances when new threats emerge. The finalized ESSEX system could also be expanded to include particle collection for detection of low vapor pressure explosives and even chemical and biological warfare agents. Its uniqueness lies in the highly selective ionization scheme and the simultaneous detection of two independent molecular properties—ionization energy and mass—which results in an extremely low false alarm rate and the unmatched adaptivity to new threat substances.

At present, systems comparable to ESSEX, developed for other purposes than explosives detection are comparable in size to present security checkpoint systems, including the input/output control stand. However, using modern components such as smaller lasers and mass spectrometers the system can be made much more compact. Thus it can easily be implemented in the current screening portals at airports and checkpoints or in other places where inspection is needed. It will be mobile and very easy to operate. It will also be possible to implement it at new places such as subway entrances, entrances to important buildings or stores. Since only vapor will be sampled it is possible to place a concealed sampling nozzle and transport the air at least a few meters to the detector which can then be placed in a separate room if this is preferred. However, the system can be made small enough to fit into such diverse places as a ticket machine or a ventilation system. A detection system that has the combination of all these features—rapid throughput and ultra high sensitivity, extremely low false alarm rate, user friendliness, advanced data processing, and the ability to target a very large number of explosives—will provide a uniquely powerful tool for security personnel at airports and other checkpoints, concealed or open.

In developing ESSEX, we will address the recommendation from the Committee on Assessment on Security Technologies for Transportation to "Explore new technologies with higher chemical specificity that are capable of detecting a wider range of explosive, chemical and biological threat materials." The necessity for such detection methods has been accentuated by the London bombings of July 7, 2005. Incorporation of just two "new" threat substances, TATP and HMTD, requires extensive research and modification of the existing hardware. With ESSEX, incorporation of new threat substances is a matter of updating the database.

6.2. Overall Methodology

The LI–MS detection method combines the two different physical principles of optical spectroscopy and mass spectrometry, giving information about two different molecule-specific properties: their mass and the energy of a molecular specific level. The REMPI technique performs the laser ionization in two steps. Figure 1 shows the simplest form of REMPI, which uses two photons of the same energy. Absorption of a first photon excites the molecule from the ground state to a molecule-specific energy level, and absorption of a second photon ionizes the molecule.

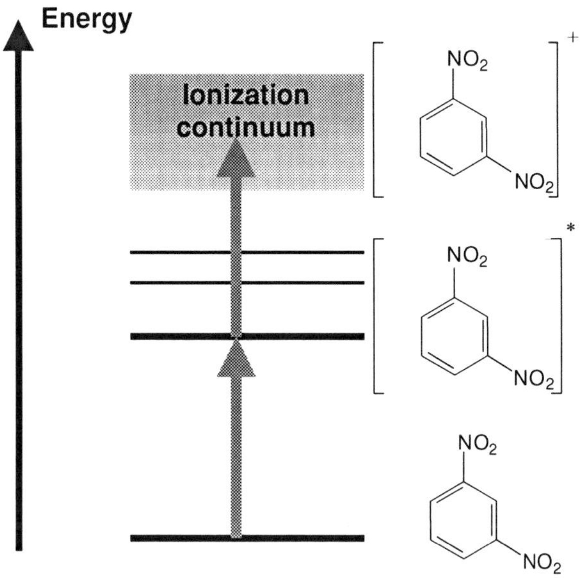

Figure 1. The REMPI process

The mass of the ion is then measured by means of a time-of-flight mass spectrometer (TOF-MS) or ion trap mass spectrometer (IT-MS), in which the ions are accelerated by an electric field and the flight time is measured. From the flight time of the ion the mass can be deduced.

For a molecule to be ionized, the energy of the laser photons must match the first excitation step. Therefore, the laser ionizes only molecules with a matching molecule-specific energy level. Since mass measurement is exclusively possible with ions, no molecules other than the targeted molecules will be detected; thus there is no interference from the surrounding environment. LI is a "soft ionization" method, which produces exclusively or predominantly the parent molecular ion; i.e., fragmentation of the ion into smaller pieces is negligible in most cases.

The REMPI technique results in a higher degree of chemical selectivity than ordinary LI due to the resonance of the first step. By proper choice of the laser wavelength for REMPI, only molecules having a level resonant at the energy of the laser photons will be ionized. This method has proven effective for the detection of one trace compound at ppt levels in the midst of others, including molecular isomers in the gas phase [3–6]. REMPI is also a very effective ionization method, which makes it highly sensitive.

A large number of explosives pose a threat now or may do so in the future. Since measurements for each substance today typically need 1 s (can be improved), it is necessary to use intelligent data processing with a combination of inputs from different sources to determine which substances to look for at each time. Such information can come from various sources. The operator can provide input (online if desired) based on the current threat situation. It is also possible to use less specific ionization methods such as single photon ionization or electron ionization to provide information about what may be present and what can definitely be excluded. This provides a threat assessment based on each detection situation and can be a basis for deciding which substances to investigate more thoroughly. The presence of other substances that are not in themselves dangerous but have higher vapor pressure and thus are easier to detect (e.g., plasticizers used to formulate military and commercial explosives) can also serve as a warning that warrants more thorough search.

6.3. TOF-MS Detection

The main advantages of the TOF-MS technique are simple and rugged design, high transmission of ions, and ability to acquire a complete mass spectrum for each laser pulse.

Since the ion yield is dependent on the laser wavelength and the mass-to-charge ratio of the ionized molecules are determined by the TOF mass spectrometer we have established a two-dimensional detection scheme based on the combination of laser absorption spectroscopy and mass spectrometry. This combination produces a high degree of selectivity at very low concentration levels on a real-time basis, which is a critical feature for the reduction of false alarm rates.

The described approach is similar to the well-established GC/MS technology. However, instead of measuring the retention time dependent Mass Spectra we exchange the retention time with the more quicker and efficient wavelength dependent laser ionization.

We envision a completely computer controlled instrument which is able to rapidly and simultaneously detect through this two-dimensional scheme as many compounds as required for any given application. A schematic drawing of the experimental setup is shown in Figure 2.

6.4. IT-MS Detection

To further improve the capabilities of LI–MS, we plan to build a prototype with an ion trap mass spectrometer instead of a TOF. IT-MS has further advantages compared to TOF-MS. First of all it is very compact enabling a very small and compact final product. Like the TOF-MS it is pulsed and can acquire a full mass spectrum for each pulse. It stores all ions from a pulse in an ion trap and

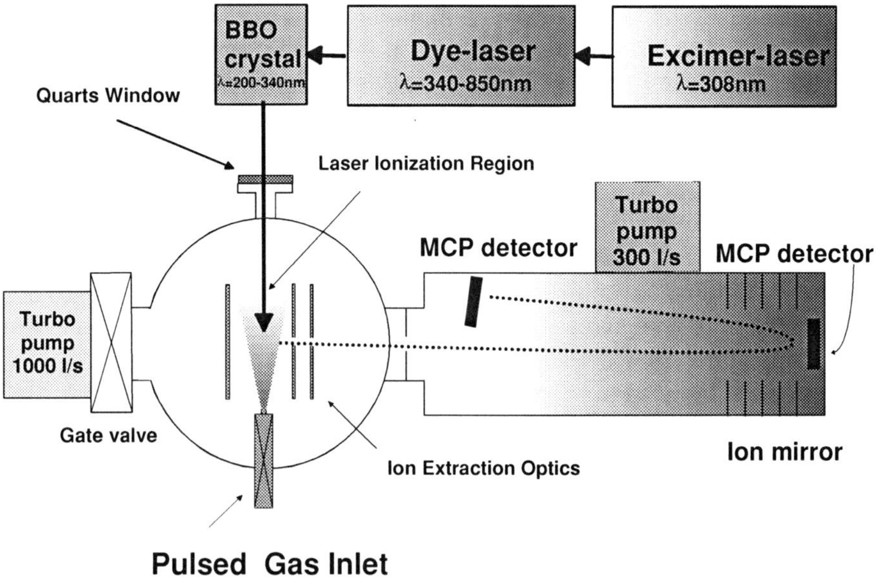

Figure 2. Setup for LI–MS detection. A laser beam from the frequency-doubled dye-laser, pumped by an Excimer laser ionizes the molecules from the sample inside a vacuum chamber. The sample gas is introduced into the chamber through a pulsed valve. The ions are extracted into the TOF-MS. At the end of the MS an ion mirror reflects the ions towards the detector

selectively ejects all ions of a certain mass. This way, like the TOF, no ions are wasted. The pulsed nature of the IT-MS makes it ideal for combination with laser ionization. Furthermore, the IT-MS has better sensitivity than a TOF-MS and it is possible to use a higher working pressure of the system. The latter means that it is possible to sample larger volumes of air with each pulse which also gives a better sensitivity. IT-MS also has capability to do MS–MS, i.e. to select a certain mass fragment for further fragmentation. This is an excellent tool for identification of a substance in cases where the selective laser ionization step fails and so provides a backup solution for difficult cases.

6.5. Physical Principles Behind Jet-REMPI–MS

With REMPI the ionization happens in two steps. The absorption of the first photon leads to the population of an excited molecule-specific level. Only the second absorbed photon results in the release of the electron. The ionization efficiency for REMPI can be extremely high, and ionization efficiencies from 1% to 10% have been reported [6].

High signals are thus produced with no physical or chemical preprocessing required. Many rovibrational levels in the excited state can be used for REMPI. The choice of level is influenced by ground state population, excited state lifetime, energy gap to the ionization potential, and disposition for fragmentation.

The sensitivity and specificity for REMPI can be dramatically improved when performed using gas cooling via a supersonic jet (jet-REMPI). In a supersonic jet, the adiabatic expansion leads to dramatic cooling, providing temperatures down to approximately 18 K [1]. This cooling greatly simplifies the spectra, producing narrower and stronger electronic transitions. The lower gas temperatures lead to population of fewer rovibrational levels, which in turn produce larger peak signals in the REMPI spectra, thus sensitivity is improved. Also, selectivity is improved because there are fewer absorption lines of other molecules that might interfere with the absorption lines of the target molecules.

Figure 3 provides an example of the improvement in ionization selectivity due to cooling. The figure shows optical spectra for toluene with cooling, where air is the carrier gas, and without cooling. With cooling, a tremendous reduction in the spectral line widths is observed.

Ions produced by laser photon ionization are ideally detected using a time-of-flight mass spectrometer (TOF-MS) or an ion trap mass spectrometer (IT-MS), which takes advantage of the pulsed nature and well-defined temporal character of laser ionization. The simultaneous detection by mass and wavelength yields a two-dimensional detection scheme (see Figure 4) based

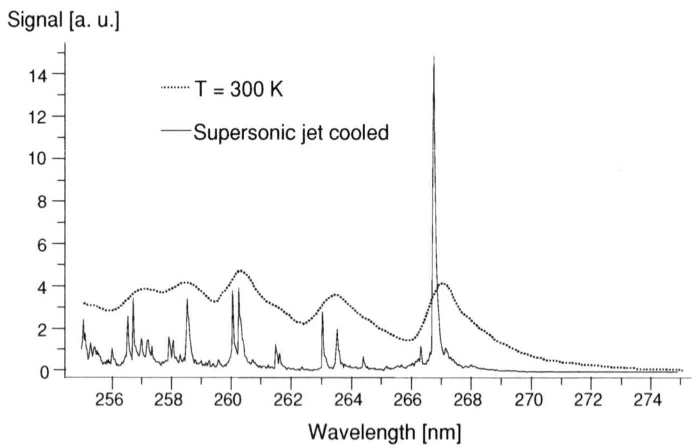

Figure 3. REMPI spectra of toluene at 300 K and supersonically jet cooled. Cooling of internal molecular degrees of freedom (rotation, vibration) in a supersonic beam results in a very narrow line allowing high-resolution gas-phase UV spectroscopy to be performed

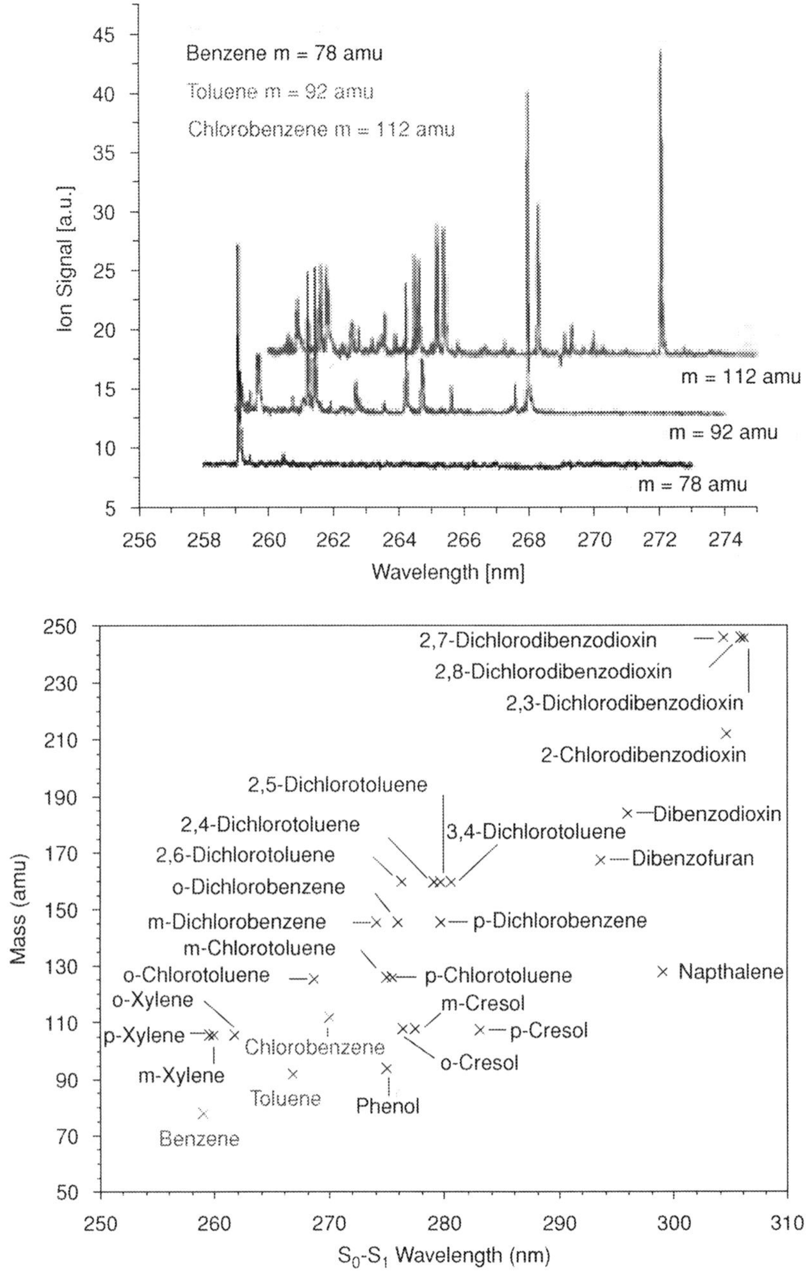

Figure 4. REMPI spectra are measured in two dimensions

on wavelength and mass. As shown in the figure, an extremely high chemical selectivity is obtained. This is crucial when identifying one trace compound in the midst of many other similar ones.

6.6. Inlet System

The sensitivity and specificity for REMPI can be dramatically improved by using cooling with a supersonic jet based inlet [8]. This cooling greatly simplifies the spectra, producing narrower and stronger electronic transitions and allows sampling directly from air. Sampling the air directly has the enormous advantage of being very fast since no extra sample preparation is necessary.

6.7. Energetic Materials as Target Molecules

Energetic materials are molecules of different classes of substances that have different physical and chemical properties with respect to each other. It is therefore likely that the optimal ionization scheme will vary with the type of explosive studied. Therefore a number of ionization schemes should be investigated. The generality of the experimental equipment allows the implementation of different schemes.

6.8. Possibilities for Data Processing and Presentation

An important feature for this type of system is incorporation and refinement of data from one or more sources. A source could be a threat assigned by the operator, a threat assigned through networked information, or the result from a sensor. The system needs to be able to handle uncertain, incomplete, and contradicting information and still perform a situation and threat analysis and present it in an easily understandable format for the operator. To have flexibility and retain the possibility of using the instrument in different situations, it will be important to make this a networked information communication and processing system, interconnecting the human operator, an intelligent computing system, and a distributed sensor system. The computing system should be able to control where the sensors scan, based on input from the human operator and the results from previous scans.

Data processing and presentation of the results are two very important factors for this type of instrument. The LI–MS detection method is very well suited for advanced data processing in several ways.

The method is adaptive and can therefore be trained for new threats. This means that the instrument can easily be updated via a CD or USB memory to detect a substance that has not been included in the database at the time of manufacture as soon as the threat has been identified and laboratory measurements have determined what parameters should be used.

The number of explosives that pose a threat now or may do so in the future is very high. Since measurements for each substance would typically need one to a few seconds it is necessary to use intelligent data processing with a combination of inputs from different sources to determine which substances to look for at each time. Such information can come from various sources. The operator can provide input based on the current threat situation. It is also possible to use less specific ionization methods such as single photon ionization or electron ionization to provide information about what may be present and what can definitely be excluded. This can be a basis for which substances to look for.

The difficulty to find a certain substance varies with (among other things) its vapor pressure. Substances with high vapor pressure are easier to find because more molecules will be present in gas phase. Some explosives have high vapor pressures and are easier to find than others with very low vapor pressures. With an intelligent data processing system, the presence of other substances that are not in them selves dangerous but have higher vapor pressure can serve as a warning that warrants a more thorough search. Such substances can for example be plasticizers used to formulate military and commercial explosives.

Depending on the input from these sources, when in doubt, a decision to make a longer measurement to look for a certain explosive can be made. Since the method is integrating, a longer measurement will give a more reliable result.

All these decisions can be made by the instrument itself without help from the operator (apart from input about the current threat). The result that the operator will see is a response of the type "yes", "no" or "maybe". In case the response is "maybe" the amount of risk should be stated and a course of action should be suggested. An example: "There is a medium risk that there is RDX present. A longer measurement time of one minute is suggested." The longer measurement time required in uncertain cases should be tolerable even at situations where a lot of people need to be screened since the time needed for some people will be much shorter.

6.9. Performance Characteristics

A fully developed ESSEX system would consist of a LI–MS detection unit and a data collection, processing and presentation unit. The system would be designed for ease of use and would require little or no user interaction. Due to the time limitations of this project, the prototype system proposed here will not have fully developed data processing and presentation but will require more skill from the operator. However, the possibilities for the system with this part developed are also described below.

It is difficult to estimate how far away from an object explosives vapor from high vapor pressure explosives are available for detection in high enough concentrations. This depends very much on the local conditions such as drafts and turbulence in the air from people walking or ventilation systems but also on the detection limit of the detection method in question. However it has been reported that TATP has been detected 2 m downwind from the source. This would be quite enough to make clever sampling in many cases, e.g. by sampling at narrow passages and providing an artificial draft in the right direction from where people pass. It is not known to us which detection method was used and what sensitivity the instrument had, but it should be noted that if a more sensitive method is used, this distance may be increased.

Unique aspects of the finalized ESSEX-system are:

- High detection selectivity. This leads to very few false positives as well as the possibility to extend the number of interrogation substances and adapt the system to emerging threats. The high selectivity makes ESSEX highly resistant to interferences from other substances and not prone to saturation. The high selectivity is due to the two different selection criteria: excitation wavelength and mass of target molecule.
- High detection sensitivity. High sensitivity makes it possible to detect a very small amount of trace residues. This may reveal a concealed explosive device or detect traces from handling of explosives on a person's hands, clothing or luggage. Estimated LOD is in the low-ppt regime. High sensitivity is a result of the resonant ionization process, giving an ionization yield as high as 10% and discriminating all contaminants of the sample.
- High throughput. An estimated average screening time of 10 s makes it possible to conduct a thorough search for threat substances on 360 persons or objects per hour. Depending on the threat situation, the operator can choose between maximizing throughput and minimizing the false alarm rate. The relatively short interrogation time results from the combination of two very fast techniques: laser ionization and time of flight mass spectrometry.
- Vapor sampling. Volatile explosives with high vapor pressures (e.g., TATP) will be sampled directly from the vapor phase, whereas for explosives with low vapor pressures (e.g., RDX), particle sampling is probably needed. ESSEX could be developed for concurrent vapor and particle sampling but for this particular project only vapor sampling will be developed.
- Advanced data processing. ESSEX can perform situation and threat analysis in circumstances of uncertain, incomplete or contradicting information. This could include decisions on where to scan and the integration time

needed to resolve the uncertainty. Input from the operator on the current threat situation is not required, but could be given. The result will be presented as a yes/no decision on whether an explosive is present. Following a yes decision, the identity of the threat substance will be given. Advanced data processing is possible due to the precise, two-dimensional data obtained upon analysis.

- Expandable, threat adaptive detection possibilities. As new threats emerge, the database can be expanded to incorporate them. Thus the ESSEX system can be upgraded and fully operative in screening for new threats in a very short period of time simply by inserting a CD or USB memory. It will also be possible to upgrade the system to detect additional substances such as chemical warfare agents, drugs, or other illicit substances. The total number of target molecules that can be included is not limited, but if a full interrogation scheme is to be performed, the interrogation time will increase with the number of target molecules. The intelligent data processing system could be used to speed interrogation by selecting the most likely target molecules based on the current threat assessment. The database is expandable due to the high specificity in both excitation wavelength and mass spectra in combination with the discrimination of background substances.
- Networked information, communication and processing. This will enable the system to take input from the operator (if desired) as well as perform near-real-time evaluation of the situation. It can thereby adjust the accuracy relative to the speed needed in the current scenario. Parameters that could be varied are interrogation time, choice of interrogation substances, and the choice between maximizing throughput and minimizing false alarm rate. The analysis can be targeted toward substances that, from other evidence, are suspected to be present. If a certain yes/no response cannot be given for a specified threat substance, the interrogation can automatically be repeated for that specific substance with an increased interrogation time.

6.10. Comparison with Existing Systems for Trace Detection

In a way, the best available method to find illicit substances by trace residues today is canine detection. Compared with technology-based trace detection systems, dogs have excellent mobility and the ability to follow a scent to its source. They can in principle be trained to detect all explosives, but there is a limit to the number of targets for which a dog can be trained (e.g., dogs used in the US military are typically trained for 9 explosives). Disadvantages with canine detection include limited duty cycle ($\sim$1 h), their need for regular training (also when not used), high costs related to the care of the dog (food, veterinary care, constant contact with the handler) and the inability to communicate the type of threat to the handler. The performance of the dog varies very much for each dog/handler team.

The most common commercially available detection method used today as a screening tool at airports is IMS, which is quite sensitive, easy to use, field deployable and available at reasonable cost. However, it has several limitations, discussed for example in "Opportunities to improve airport passenger screening with mass spectrometry" by the Committee on Assessment of Security Technologies for Transportation [2]. Two of the limitations are the relatively low chemical specificity, resulting in a higher level of false alarms, and the limited number of explosives concurrently detectable. A recommendation from the committee was to "Explore new technologies with higher chemical specificity that are capable of detecting a wider range of explosive, chemical and biological threat materials." The committee also expected MS-based systems to be capable of alarm thresholds 1000 times lower than IMS systems due to the higher chemical specificity. The context of this statement was airport security but it holds equally true for other applications such as detection of suicide bombers.

Some detection methods are able to give fast response for one single substance. One such method is surface acoustic waves (SAW), in which a specific selective coating is used to achieve chemical specificity. The selectivity is very good, but the range of compounds detectable is very limited and only useful for very specific tasks.

A common way to address the low specificity of detection methods is to add a GC step to achieve temporal discrimination of signals from different molecules. This can be used in conjunction with IMS, SAW, MS, and other methods. The longer the GC column, the more separation is achieved but also the longer it takes. Therefore, there is a trade-off between increasing the specificity and maintaining a high throughput.

Table 1 compares current explosives detection technologies. It should be noted that the methods selected here are only examples. Among commercially available systems are many combination methods as well as completely different methods that are not mentioned here.

The ESSEX detector has the advantage of being very selective due to the combination of resonant laser ionization and mass spectrometry. The resonant laser ionization serves as a preselection step for the mass spectrometric measurement, much as GC is a preselection step for many methods. The difference is the speed with which ESSEX can make this step compared with, for example, GC. The detection limit is expected to be very low (ppt range) for most substances, and the high selectivity should allow for a low alarm threshold compared with IMS, which helps maintain a low effective detection limit. The low chemical specificity of IMS systems makes them susceptible to interferences from many substances present in the surrounding environment

TABLE 1. A comparison of some currently existing explosives detection technologies. Among commercially available systems are many combination methods

Attribute	ESSEX	Dogs	IMS	GC–MS	SAW/GC–SAW
Measurement time	Fast	Fast	Fast	Slow	Slow
Throughput rate	High	Low	High	Low	Low
Detection limits	ppt/ppb	<ppb	ppt	ppt	ppb
Molecular selectivity	Best		Poor	Best	Good/fair
Range of compounds	Good expandable	Good trainable response	Poor	Good	Poor/good
Number of compounds	Excellent	9	6–10	Excellent	1/good
Typical operational conditions	No consumables	High consumables	Radioactive ionization source	High consumables	Apparatus dependent
Application remarks	REMPI in pilot/field use	Capabilities decline with time; fatigue	Available as hand-held system for civilian and military (CAMS)	Field systems available	Very small, robust. Basis for JCAD (joint chemical agent detector)

such as perfumes and hand creams. This can sometimes be exploited to suppress a positive result, thus avoiding detection of concealed explosives. Due to the higher specificity and discriminating ionization process of the ESSEX detector, the problem with interferences and false negatives from suppression will be avoided.

6.11. Future Potential

The potential for further development of this technique is excellent. The detection method itself is very general and applicable to a wide range of substances. This means that the database could be expanded to include drugs and chemical agents.

Many explosives have vapor pressures that are too low to be present in air in sufficient quantities even for extremely sensitive detection methods. Nevertheless, dogs can often detect explosives at these conditions. Although their method of collection is not fully understood, the current hypothesis is that dogs inhale either dust particles with explosives adsorbed on the surface or small explosives particles along with the explosives vapor. Therefore, for a trace detection system to work for low vapor pressure explosives, appropriate sampling of particles is necessary. Even though particle sampling systems are available for combination with other detection methods, they can probably not be used unaltered with ESSEX.

Technical development allowing fast changes of wavelength (i.e., change of target molecule) with good accuracy and higher laser frequency will result in a faster system. The limiting factor will then be the time necessary for particle collection. Increasing the scan speed and repetition rate of the laser will decrease the detection time and thereby the throughput rate. Alternatively, more substances can be targeted in each interrogation or the integration time can be increased to further lower the signal-to-noise ratio (S/N). If the repetition rate of the laser can be increased by a factor of 100, the S/N will improve by a factor of 10.

Better ion optics and detectors can improve the sensitivity of Jet-REMPI systems with up to 100 times compared to systems with parallel plate ion extraction and MCP detectors [9]. Such improvement would enable detection further from an object or detection of explosives with lower vapor pressures at shorter distances.

References

1. Oser, H., M. Coggiola, J, G.W. Faris, S.E. Young, B. Volquardsen, and D.R. Crosley, 2001. Development of a jet-REMPI (resonantly enhanced multiphoton ionization) continuous monitor for environmental applications, Appl. Opt. 40 (6), 859–865.
2. Committee on Assessment of Security Technologies for Transportation, National Research Council of the National Academies, 2003. Opportunities to improve airport passenger screening with mass spectrometry, The National Academies Press, Washington, DC.
3. Oser, H., K. Copic, M. Coggiola, J, G.W. Faris, and D.R. Crosley, 2001. Congener-specific detection of dioxins using jet-REMPI. Chemosphere, 43 (4–7), 469–477.
4. Thanner, R., H. Oser, and H.-H. Grotheer, 1998. Time-resolved monitoring of aromatic compounds in an experimental incinerator using an improved jet-resonance-enhanced multi-photon ionization system Jet-REMPI, Eur. Mass Spectrom. 4 (3), 215–221.
5. Zimmermann, R., H.J. Heger, M. Blumenstock, R. Dorfner, K.W. Schramm, U. Boesl, and A. Kettrup, 1999. On-line measurement of chlorobenzene in waste incineration flue gas as a surrogate for the emission of polychlorinated dibenzo-*p*-dioxins/furans (I-TEQ) using mobile resonance laser ionization time-of-flight mass spectrometry, Rapid Commun. Mass Spectrom. 13 (5), 307–314.
6. Lubman, D.M., and L. Li, 1990. In Lasers and Mass Spectrometry, edited by D.M. Lubman, Oxford University Press, New York.
7. Boesl, U., 2000. Laser mass spectrometry for environmental and industrial chemical trace analysis, J. Mass Spectrom. 35, 289–304.
8. Imasaka, T., D.S. Moore, and T. Vo-Dinh, 2003. Critical assessment: use of supersonic jet spectrometry for complex mixture analysis, Pure Appl. Chem. 75 (7), 975–998.
9. Suzuki, T., S. Hayashi, S. Ishiuchi, M. Saeki, and M. Fujii, 2005. A new, highly sensitive, time-of-flight mass spectrometer consisting of a Flangeon-type conical ion lens system and a prototype daly detector for exhaust gas analysis based on the Jet-REMPI technique, Anal. Sci. 21, 991–996.

7. HIGH SPEED MID INFRARED DETECTION OF EXPLOSIVES USING LASER SPECTROSCOPY

A. Hofmann and A. Lambrecht
Fraunhofer Institute for Physical Measurement Technology IPM, Heidenhofstrasse 8, D-79110 Freiburg, Germany

Abstract. A prototype of a laser spectroscopic gas detection system will be developed which allows high speed detection of explosives around a person or a mobile object up to distances of 1 m. High speed detection is a prerequisite for moving targets (e.g. on a conveyor belt) or for fast screening of persons and/or objects. A vision of such an instrument could be similar to a vacuum cleaner.

Key words: middle infrared, explosives detection, laser spectroscopy, high speed sensing

7.1. Introduction

First activities of the Fraunhofer Institute of Physical Measurement Techniques (IPM) within the context of the proposed project are described. A project plan including a description of the detection device and several application scenarios is also presented.

7.1.1. BACKGROUND IN OPTICAL SYSTEMS

Fraunhofer IPM has been working for several years in developing and assembling complex optical measuring devices as well as constructing laser gas spectroscopy systems.

Several systems developed by Fraunhofer IPM prove the institut's competence in establishing complex, reliable, robust and accurate optical systems, for example:

The cine film recorder "Arrilaser", a laser imaging system which records digital or digitised pictures on cinema projection film, has been awarded the "Scientific and Engineering Award" by the Academy of Motion Pictures Arts and Science in 2002. The system triggers three lasers simultaneously at a high speed of 30 ns achieving an accuracy within per mille. High speed, high precision and high resolution are the basis parameter of its commercial success.

H. Schubert and A. Rimski-Korsakov (eds.), Stand-off Detection of Suicide Bombers and Mobile Subjects, 51–58.

The Microscopic Particle Monitor, automatically reconstructs a 3D-picture of trapped bio aerosols. These aerosols are counted, analysed and classified according to their species independent of their actual spatial position in the trapping membrane. The membrane is automatically exchanged several times per day and analyzed inside the monitor system. The system distinguishes itself by its robust design and its autonomous operation even under extreme weather conditions.

Reliable operation regardless of weather conditions is a major feature of Fraunhofer IPM's railroad measurement systems, too. Attached in front or on top of a measurement train driving at velocities of about 200 km/h, IPM's clearance profile scanner or contact wires wear positioning and degree measuring systems are running in over 20 countries worldwide. Laser scanning technology is used and achieves data acquisition rates of 1 MHz with a 2D-positioning accuracy of 0.1%. Reliable operation is guaranteed at ambient temperatures from $-20\ ^{\circ}C$ to $+50\ ^{\circ}C$.

7.1.2. EXPERIENCE IN LASER GAS SPECTROSCOPY

In the area of laser spectroscopy Fraunhofer IPM has shown its expertise in several measurement devices for a wide range of applications. The highest accuracy and selectivity for molecules in the gas phase is achieved by MIR laser systems, as the fundamental vibrational band of most molecules is in the MIR range. Fraunhofer IPM has developed MIR-laser based gas detection systems which are able to detect various industrial or traffic trace gases over distances of several meters. These measurement devices are equipped with Quantum Cascade Lasers (QCL). QCL are as small as visible semiconductor lasers and can be run at ambient temperature in pulse mode. If a QCL emitting in the spectral absorption band of the gas of interest is available, a suitable source for an accurate and selective gas measurement system can be build. Fraunhofer IPM has already assembled and tested a compact laser MIR spectrometer, which can detect small gas concentrations over distances of about 20 m. The measured detection limits, normalized to distance of 1 m and an integration time of 1 s, were 54 ppb for NH_3 (at 10.33 μm), 189 ppb for CO_2 (at 4.3 μm) and 1.9 ppm for NO (at 5.45 μm). The system has shown a good long term stability [1]. In Figure 1 measurement results using NO gas with concentrations up to 150 ppm are shown. As the clip shows in more detail the gas concentration was changed in defined steps.

Similar measurements were done with CO. The over all drift was measured to be 1.3 ppm over 8 h at a concentration level of about 40 ppm. The measurements are also described in detail in [1].

In addition to long-term stability tests fast gas measurements at extremely dynamic gas exchange rates are performed. Fraunhofer IPM developed and

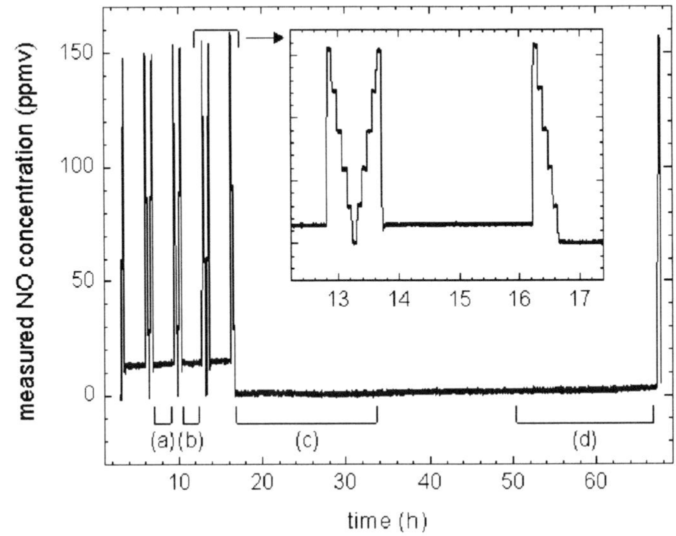

Figure 1. Stability of MIR laser spectroscopy system measuring NO gas concentration. A statistical data analysis was performed in the indicated sections (a), (b), (c) and (d)

constructed a Dynamic Exhaust Gas Analyzing System (DEGAS) for measuring the dynamic gas change rates of an automobile engine running at several thousand rotations per minute. The DEGAS system is developed for engine test beds equipped with twelve measuring channels measuring three gas components each at four different locations. The time resolution of the DEGAS system is 5 ms which allows distinguishing the gas exchange in the exhaust area. This fast measurement rate is due to a very effective probe sampling and commutation device. In [2] the system and measurements are described.

Fraunhofer IPM has also realized a small and fast gas measuring unit measuring the CO_2 for breath testing. Using a VCSEL laser at 2 μm a time resolutions of less than 1 ms was achieved. The dimension of the device could be reduced to a convenient handheld, battery driven system. In Figure 2 results of VCSEL stability and accuracy measurements of CO_2 are displayed. From the CO_2 data a noise equivalent concentration (NEC) of 0.1% CO_2 can be determined [3].

7.2. Proposal for a Fast Explosive Detection System Using MIR Laser Spectroscopy

For the real-time detection of explosives a fast analysis of the signals is crucial. We envisage a measurement system which evaluates real-time data in a time scale of about 10 ms. The basis of the method is laser spectroscopy in the mid

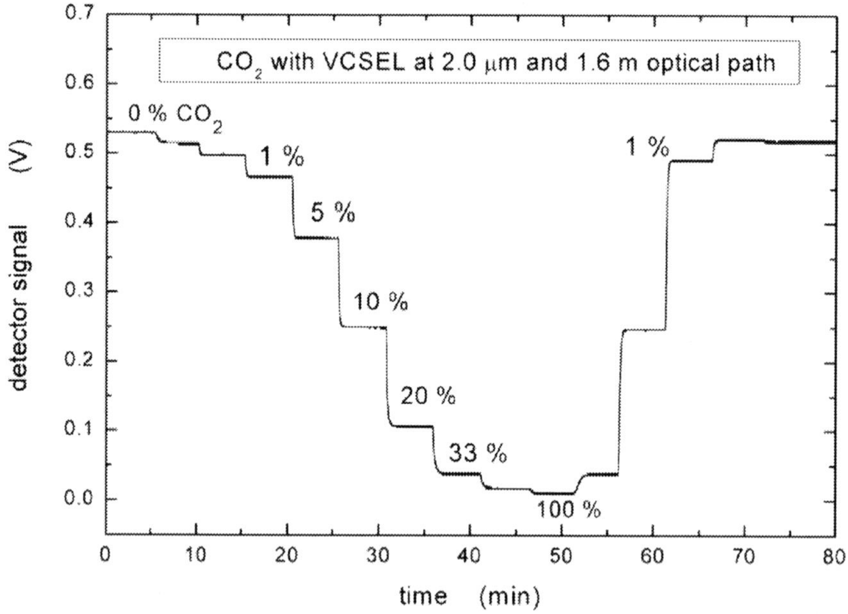

Figure 2. CO_2 measurement using a VCSEL source at 2.0 μm

infrared (MIR) where highly selective and sensitive detection of gases can be performed. The ultimate time resolution of such a system is limited by the time to scan a single rovibrational absorption line of a molecular gas. Using pulsed QCL this can be done in less than a microsecond. The system will be used as non-contact sampling unit. Depending on the scenario a stationary unit sampling up to 1 m or a transportable unit working at distances of about 10 cm can be realized. The sales price of the system should not exceed 40,000 ° in order to guarantee its prevalence. This will be achieved using a modular design.

The complete system consists of sampling unit, detection unit and analyzing software, shown in Figure 3.

7.2.1. SAMPLING UNIT

We propose a sampling system similar to a vacuum cleaner complemented by a pulsed air blowing. The functional details are shown in Figure 4. Naturally, the sampling distance is limited by the power of the airflow, especially for a mobile solution. However, sampling generally is faster, more sensitive, has a lower false alarm rate than non-sampling methods. Furthermore, it could be realized at lower cost. By laboratory tests using powerful equipment we shall investigate whether sampling distances up to 1 m are feasible. For the mobile

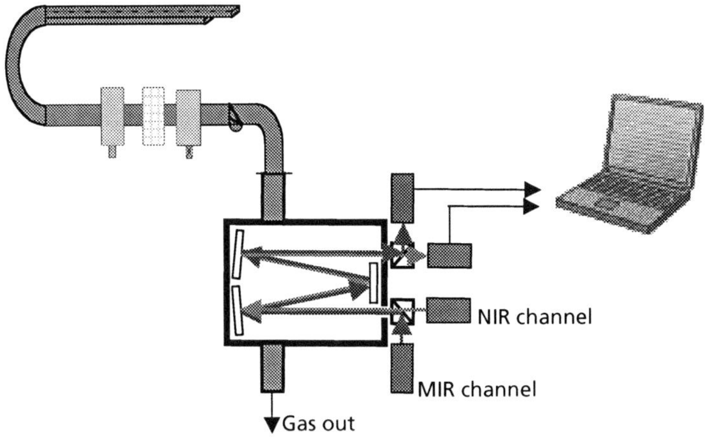

Figure 3. Schematic drawing of system design composed of sampling unit, detection unit and analyzing tools

prototype "Speedmex-vacuum cleaner" a compromise between sampling distance and system size/power will be established.

Practically the time resolution is limited by the gas sampling system. The sampled dusty air will be prefiltered with a cyclone to remove heavy objects. An additional filter will guarantee that no objects bigger than several 100 μm will pass the filter. Particles smaller than some 100 μm will be able to pass and subsequently pyrolyzed. The pyrolysis of the small particles will be done in a catalyst coated ceramic membrane which is heated up to 1000°C. A soot trap next to it will catch most of the remaining solid leaving only gas to pass on to the optical detection unit. In the optical detection unit quite constant environmental conditions with a slight low pressure of 500 mbar and temperatures of about 40–90 °C will prevail.

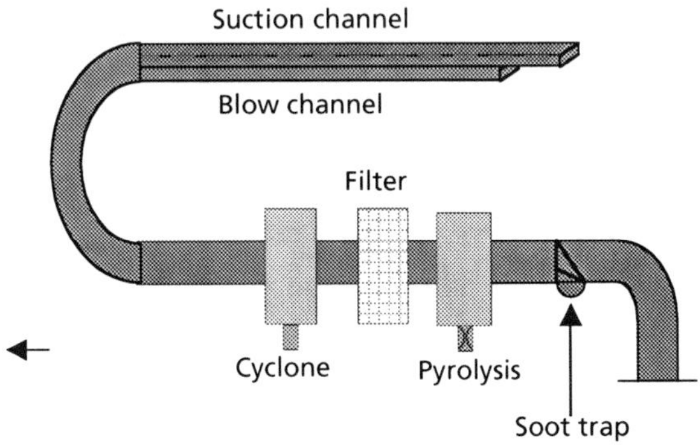

Figure 4. Functional details of the sampling unit

The resulting gases will be guided into an optimized low volume long optical path gas (White) cell. An overall measuring rate in the 10 ms range is feasible. This has already been proved by IPM for automobile exhaust analysis. All parts of the sampling unit can be demounted easily for a fast cleaning service on site.

7.2.2. OPTICAL DETECTION UNIT

We shall concentrate our measurements on NO as a special chemical substance/marker typical for explosives. Fast NO measurement using QCL was already demonstrated at Fraunhofer IPM. NO will be generated by pyrolysis with a suitable catalyst material (e.g. CuO/Pt) at temperatures in the 800–1000°C range. Additionally we shall employ a fast CO_2 measurement. Simultaneously normalizing the NO to the CO_2 signal of the same sample volume in the cell is a significant advantage of the proposed method. Most dust particles give CO_2 signals. The intensity of the CO_2 signal depends on the particle size. Thus, the false alarm rate is expected to be greatly reduced by normalization (see Figure 5).

NO is determined using a pulsed QCL at approx. 5.3 μm, CO_2 is determined with a surface emitting laser (VCSEL) emitting at 2.0 μm in cw mode. Both detection techniques can be combined with the same sample cell and measurement rates below 10 ms are feasible. NO can be detected to a

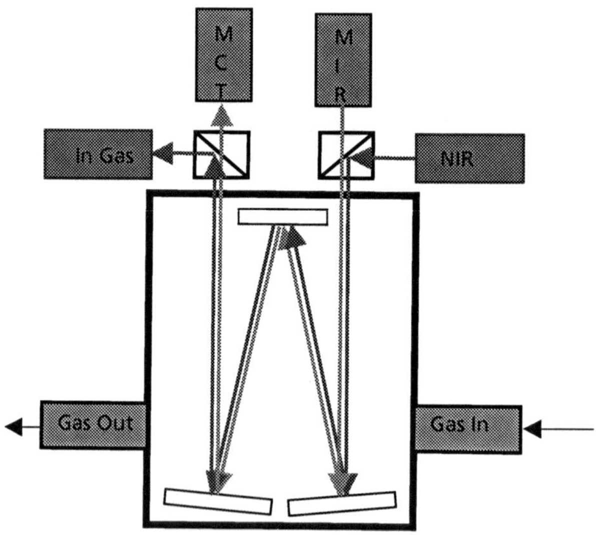

Figure 5. Functional details of the optical detection unit

resolution of 1 ppm, CO_2 can be detected down to 10 ppm. The detection units — MCT detector for the MIR and InGaAs detector for the NIR — can be driven at room temperature. The optical setup will be modular; therefore an easy exchange of the QCL can be accomplished on demand in cases where alternative wavelengths have to be measured. This setup is based on a compact White cell which is described in [4].

7.2.3. SYSTEM CONCEPT AND PROPOSED SPECIFICATIONS

The system is designed as modular system usable as a stationary system for probe distances up to 1 m, as a transportable system in a vacuum cleaner design or even as portable system for an unobserved use in crowds as a backpack design. The system will achieve a high sensitivity in the 1 ppm range for NO detection. The detection speed will be 100 ms per measurement resulting in real-time observation in fast changing surroundings. It will achieve high selectivity, as there are no cross sensitivities between different pollutants due to the dual wavelength measurement of NO and CO_2. The sales price should be around 40,000 ° allowing a common use of the detection device. A development time of two years for a prototype device is feasible.

Compared to other solutions available on the market this system will feature high sensitivity and selectivity, high detection speed, contactless detection and no need for disposals.

7.2.4. OUTLOOK

Several application scenarios for the envisaged explosive detection system are conceivable due to its modular construction. Two possible applications are:

- A stationary detection system for permanent monitoring of packages at airports, harbors or at customs could be installed next to a conveyor belt. Non-contact detection allows for an automated observation of several packages of different size within a time span of one second is feasible. The detection unit can be adapted to the existing cycle.
- A transportable detection unit could be used for the observation of persons, e.g. at the entrance of big public events. The operator will carry the detection unit on his back, the air intake can be realized as a short flexible tube or may be hidden in the operator's cloths. As we are dealing with a non-contact detection system, the operator who moves through the crowd remains unsuspicious. The operator moseys between the people and can relate explosives to passing individuals due to the fast detection. Identifying dangerous individuals before they enter a public event is therefore possible.

References

[1] Lambrecht, A., 2005. Quantum cascade lasers, systems and applications in Europe, Proc. SPIE, 5732, 122–133.

[2] Lambrecht, A., M. Braun, S. Hartwig, J. Nurnus, J. Wöllenstein, F. Weik, and J.W. Tomm, 2006. New infrared sources for breath analysis, Proc SPIE, 6093, 60930B.

[3] Grisar, R., W.J. Riedel, M. Tacke, H. Wolf, and E. Wagner, 1998. Zeitlich hochaufgelöste Gasanalytik: Anwendungen der Infrarotdioden-Laserspektroskopie am Beispiel von Verbrennungsprozessen, Technisches Messen, 65, 412 (in German).

[4] Hartwig, S., J. Hildebrand, M. Moreno, J. Fonollosa, J. Santander, R. Rubio, C. Cane, A. Lambrecht, and J. Wöllenstein, 2005. A highly sensitive IR-optical sensor for ethylene monitoring, Proc. SPIE, 5836, 452–460.

8. REMOTE DETECTION OF EXPLOSIVES BY ENHANCED PULSED LASER PHOTODISSOCIATION/LASER-INDUCED FLUORESCENCE METHOD

Talya Arusi-Parpar[1] and Izhak Levy
Atmospheric Optics Group, Electro-Optics Division,
Soreq NRC, Yavne 81800, Israel

Abstract. A unique scheme for remote detection of explosives has been demonstrated. High detection sensitivity was achieved by an exclusive combination of two processes: Explosive vapor detection by PLP/LIF (pulsed laser photodissociation/laser-induced fluorescence) with vapor concentration enhancement by remote evaporation. The basic PLP/LIF method was demonstrated on TNT in the vapor phase with a detection sensitivity higher than 1 ppb m. Remote vapor enhancement improves detection capability by about three orders of magnitude enabling remote detection of explosives with extremely low vapor pressures. Remote detection of RDX, PETN and C4 from several meters is presented. Updated experimental results and calculations show that the enhanced PLP/LIF method is a promising tool for detection distances of up to ten's of meters.

Key words: explosives, photodissociation, remote detection, laser-induced fluorescence

8.1. Background

Most of the existing high sensitive methods for detection of explosives involve sampling of air or traces around the suspected area, and analyzing it by different means [1–5].

However, these methods are time consuming and require physical approach to the suspected area. Other conventional methods based on X-ray, gamma ray detection and nuclear resonance absorption [3,4,6] cannot be applied as remote sensing tools. Furthermore, there is a need for sensitive remote detection of explosives. The demands from such methods require detection of trace levels of explosives in ambient conditions, *in situ* and in *real-time*.

[1] E-mail: parpar@soreq.gov.il

H. Schubert and A. Rimski-Korsakov (eds.), Stand-off Detection of Suicide Bombers and Mobile Subjects, 59–68.

Over the last few years optical techniques, particularly those based on laser spectroscopy, were intensively investigated in order to develop a highly sensitive optical explosives detector. The use of laser technologies for the detection and identification of organic molecules and explosives has been utilized by a variety of approaches. Methods such as resonance enhanced multiphoton ionization (REMPI) spectroscopy [2], laser surface photofragmentation [7] or vapor detection by mid-infrared absorption in special absorption cells [8, 9], are ultra sensitive detection methods, widely utilized for the close range detection of dangerous molecules or pollutants. For remote detection, Raman spectroscopy [10, 11], and infrared vapor absorption have been extensively considered [12]. However, those methods are generally not sensitive enough for detection of explosives at low concentration levels because of the relative low cross-sections of these processes. Few are the approaches, such as laser-induced breakdown spectroscopy (LIBS) [13] and pulsed laser photodissociation/laser-induced fluorescence (PLP/LIF) [14–16] which combine high sensitivity with the option to be extended to a remote detection tool [17–21]. Our latest experimental results, updated as to March, 2006, show the ability of highly sensitive remote detection of nitro-based explosives at ambient conditions by the use of enhanced PLP/LIF method.

8.2. PLP/LIF Detection Method

Most of the high explosives are nitro-based with similar molecular structure with vapor pressures in the range of ppb to sub-ppt at room temperature. Remote detection of such extreme low vapor concentrations is a very challenging task. The basic detection concept demonstrated at NRC Soreq consists of two progressive steps produced by the same laser beam pulse. The first step is the photodissociation of the explosive molecule, whose photofragments include vibrationally excited NO radicals. The second step is the detection of vibrationally excited NO photofragments utilizing laser-induced fluorescence (LIF) method. Figure 1 describes the relevant electronic and vibrational levels for NO showing the excitation from the second excited vibrational level in the electronic ground state to the lowest vibrational level in the first excited electronic level and the relevant fluorescence transitions back to the electronic ground state vibrational levels. This method relies on the production of energetic NO radicals during the photodissociation process, therefore when detecting different nitro-based explosives, the same fluorescence wavelength will be collected, detecting but not identifying the specific explosive.

There are two important advantages in detecting the vibrationally excited NO radicals over detecting the ground state NO produced by the

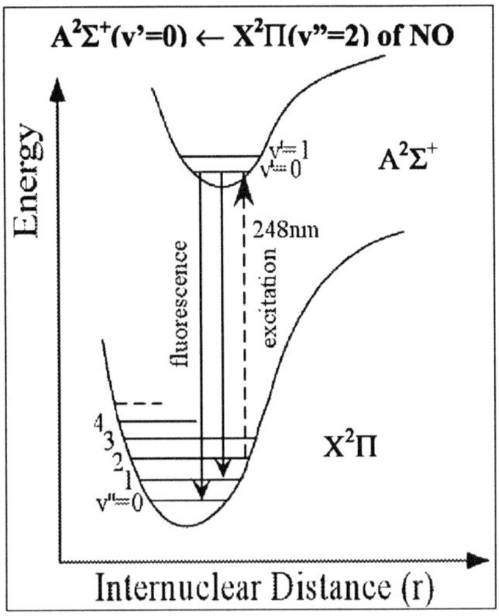

Figure 1. Ground and first excited electronic levels of NO describing the relevent detection scheme by laser-induced fluorescence

photodissociation. The first lies in the fact that the collected fluorescence is at shorter wavelengths than the exciting laser in contrast to the fluorescence occurring from either the explosive molecules or other molecules existing in air. Second, there is no background fluorescence from ground state NO present naturally in air. On the other hand the main difficulty arises due to working at ambient conditions by the dramatic shortening of the NO fluorescence lifetime. At atmospheric air pressure the fluorescence lifetime of the NO is just $\sim$1.5 ns due to strong quenching caused mainly by oxygen.

We chose to detect NO produced from the photo-fragmentation of the explosives in the second vibrational level over that produced to the first, in order to reduce the possibility of detecting thermally vibrationally excited background NO radicals, which are present in air, thus reducing the probability of false alarm.

The basic PLP/LIF method was demonstrated on TNT vapor from 2.5 m at ambient conditions [15, 16]. The equilibrium vapor pressure of TNT at room temperature is about 10 ppb. Therefore, with the basic PLP/LIF method, a sensitivity higher than 1 ppb m was demonstrated. Even though, this method is applicable for all explosives containing functional nitro groups, much higher detection sensitivities are required to detect explosives such as RDX (cyclotrimethylene trinitramine), PETN (pentaerythritol tetranitrate), SEMTEX (an RDX and PETN mixture with additional plasticizer), etc. due to their much lower vapor pressures.

8.3. PLP/LIF Experimental Setup

The experimental setup included a Spectra-Physics Lasers system. The system worked at 10 Hz repetition rate with output energy of 5–10 mJ at the required UV wavelengths (~248 nm), with a spectral linewidth of 0.1 cm^{-1} (FWHM) and a pulse length of 10 ns (FWHM). The unfocused, 4 mm diameter laser beam, was introduced to the explosive interaction region after passing through some mirrors and folding prisms. A small amount of solid explosive material (several mg) was positioned on a microscope slide about 1 cm below the beam path in the open laboratory environment.

The optical detection system was positioned 20° relative to the beam path at a distance of 250 cm from the interaction region. It included a 20 cm diameter, 50 cm focal length quartz lens for fluorescence collection followed by collimation optics to enable parallel beam path through the interference filters which reject the scattering of laser light while transmitting the fluorescence. By the use of another lens the fluorescence is focused on a solar blind photomultiplier tube (PMT, Hamamatsu R7154). The signal, collected by the photomultiplier, is transferred via a preamplifier (Hamamatsu C5594 with a frequency bandwidth of up to 1.5 GHz) to an analog integrator (BOXCAR—Stanford Research Systems, SR 250), where it was integrated during a ~15 ns sampling gate. The BOXCAR's sampling gate was positioned temporally on the laser pulse since the fluorescence of the NO overlaps temporally with the laser pulse. The signal and the sampling gate were monitored on a fast digital oscilloscope (1 GHz Tektronics, TDS 684A), while the data was collected and analyzed by a computer equipped with an analog to digital converter card (A/D—National Instruments, PCI MIO-16E-4). A National Instruments LabView program controlled the MOPO wavelength scanning range and scanning rate. This program was also used for data acquisition and analysis.

8.4. PLP/LIF Experimental Results

The TNT experiments included several scans of the $v'' = 2$ bandhead in the range of 247.92–247.98 nm. The spectra presented here include a minor wavelength calibration correction compared to the ones presented in our previous papers [15,16]. The scans included the overlapping rotational lines of the edge of the P_{12} bandhead (online) and part of the wavelength region at longer wavelengths (offline). This part of the spectrum is chosen because it is easy to identify within a small wavelength tuning range the presence of explosives by comparing the signal intensity online and offline.

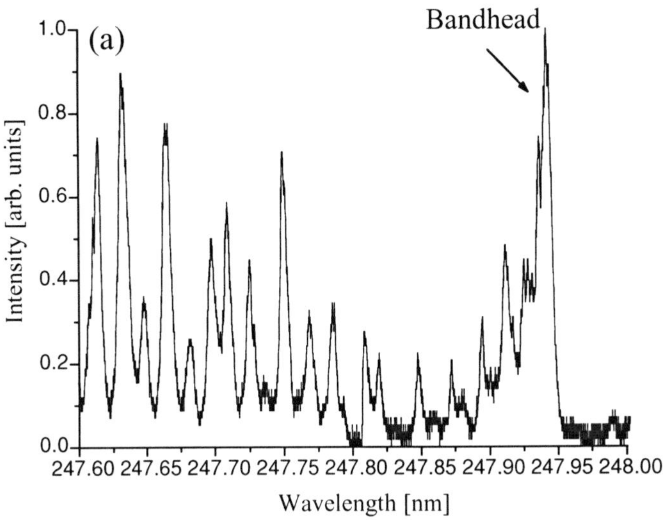

Figure 2. Part of the $A^2\Sigma^+(v' = 0) \leftarrow X^2\Pi(v'' = 2)$ transition of NO resulting from the photodissociation of nitrobenzene in vacuum

Figure 2 shows part of the $A^2\Sigma^+(v' = 0) \leftarrow X^2\Pi(v'' = 2)$ transition of NO, taken from our preliminary results of the photodissociation of Nitrobenzene in vacuum conditions.

A magnified view of the bandhead is presented in Figure 3, including the *online* and *offline* regions, which were repetitively scanned during the TNT detection experiments.

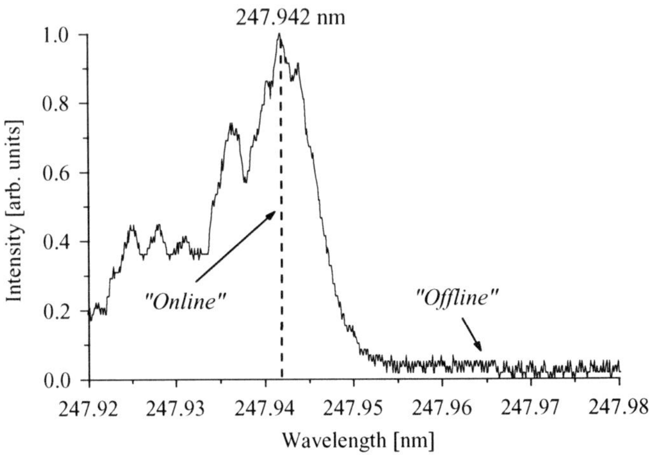

Figure 3. The NO $A^2\Sigma^+(v' = 0) \leftarrow X^2\Pi(v'' = 2)$ transition bandhead (magnified view of the bandhead in Figure 2)

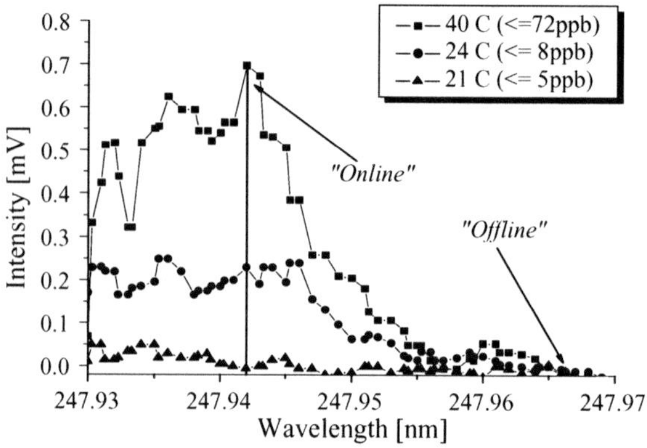

Figure 4. NO spectra resulting from TNT vapor for different temperatures (vapor pressures) detected from 2.5 m

Figure 4 shows the measured NO spectra obtained from TNT dissociation in atmospheric background pressure for different sample temperatures (different TNT vapor pressures). Each point in the spectrum is a result of an average of 1000 points, meaning 100 s. In order to increase the sample concentration we used a heating tape to control the temperature of the sample plate. It can be seen (Figure 4) that the fluorescence signal is increasing dramatically with temperature. We had no means of measuring the exact vapor concentration at the interaction zone, so the calculated vapor pressure as a function of temperature [22] was taken as an upper limit to the actual TNT partial pressure. It is known [23] that the probed TNT concentration was actually much lower than the calculated equilibrium vapor pressure. Thus, the detection sensitivity we claim to have measured is in fact higher than named. No NO fluorescence is measured when there is no TNT on the sample plate. Therefore, the natural NO or NO_2 in the air is not contributing to the NO signal measured from TNT.

Keeping in mind that the calculated vapor pressure at this temperature is an upper limit to the detection sensitivity of our apparatus, the results in Figure 4 demonstrate that we were able to detect less than 8 ppb of TNT in atmospheric pressure background at of 24 °C with a signal to noise ratio of about 10.

8.5. Enhanced Results

The basic PLP/LIF method showed encouraging results; still a much higher sensitivity is necessary to address other explosives with room temperature vapor pressures in the ppt region (RDX, PETN, C4,...). Lately we introduced a method to remotely evaporate the explosive sample thus creating

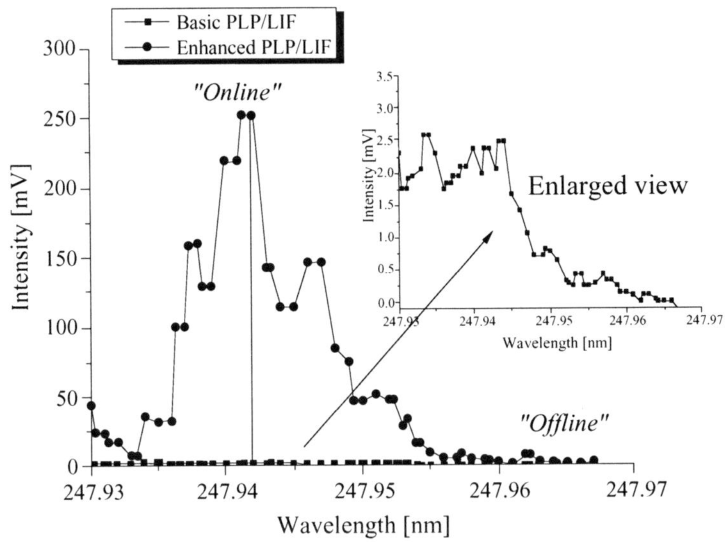

Figure 5. Spectra resulting from TNT vapor by using the enhanced PLP/LIF method compared to the previous results at the same temperature. A magnified view is given for the spectra obtained without remote evaporation

much higher concentrations of the explosive material. Experiments on TNT showed an enhancement of two to three orders of magnitude in signal with a high signal to noise ratio and a much shorter averaging time (10 s per point) opening the possibility of the detection of additional high explosives. In all the enhanced PLP/LIF experiments, explosive samples of several milligrams down to several hundreds of micrograms were positioned on a microscope slide in the open environment of the laboratory and at room temperature with no further heating as done before.

Figure 5 shows the comparison of the fluorescence spectra taken from TNT vapor with and without remote evaporation at a distance of 2.5 m at room temperature. It can be seen that an enhancement of more than two orders of magnitude can be achieved using a much shorter averaging time. A magnified graph of the previous results is shown within Figure 5. Furthermore, various nitro-based explosives were tested DNT, TNT, RDX and PETN in crystalline form and also with plasticizers, C4, Composition B and SEMTEX. For all mentioned explosives successful remote detection was demonstrated. Preliminary experiments, with other explosives such as AN (ammonium nitrate), ANFO (ammonium nitrate fuel oil) and UN (urea nitrate) showed also promising results.

Figure 6 shows some representative spectra for the remote detection of DNT, crystalline RDX, C4 and crystalline PETN from 2.5 m. For all spectra we used an average of 100 pulses per wavelength (10 s). With an operational

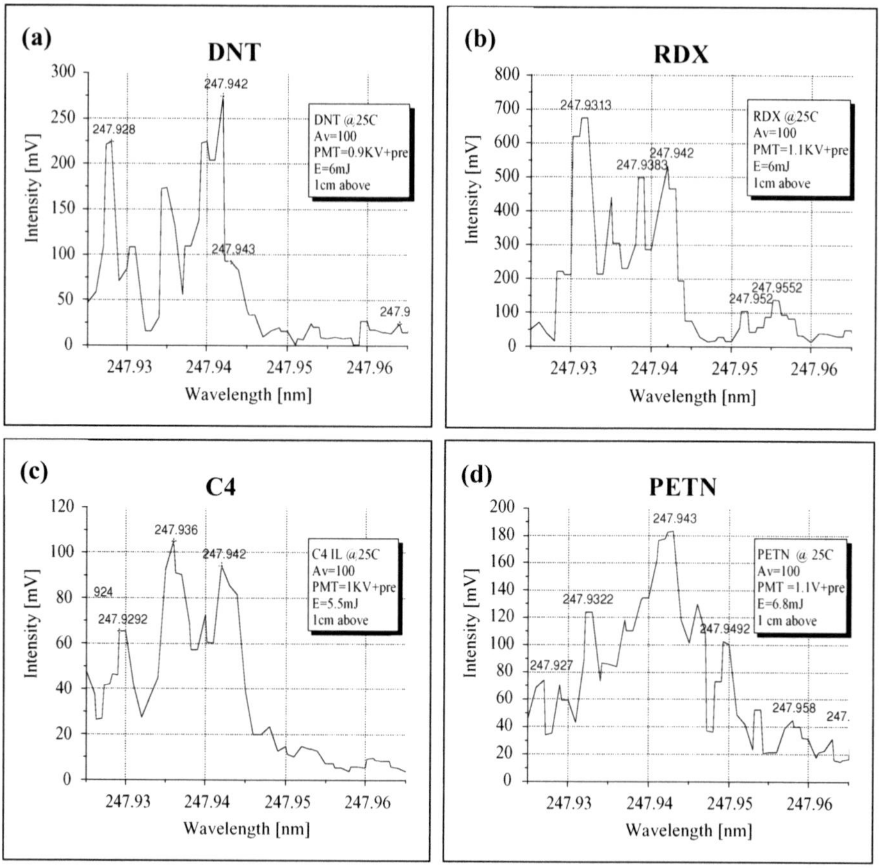

Figure 6. Spectra resulting from (a) DNT—Di-nitrotoluene, (b) crystalline RDX, (c) C4 and (d) crystalline PETN

system there is no need to scan the whole bandhead, since by probing discrete wavelengths belonging to the *online* and *offline* regions one can indicate the presence of explosive molecules (within detection sensitivity), reducing significantly the time needed for the detection. With our current system response times of two up to ten seconds were measured. The major reason for this response time is due to strong variations in vapor concentration above the explosive sample indicating that the evaporation process is of turbulent nature. In order to reduce system response time, an optimal evaporation scheme has to be obtained. Therefore further investigation is needed especially in understanding the explosive traces evaporation mechanism and the vapor concentration buildup above the explosive traces.

Preliminary experiments and calculations predict the capability of increased detection ranges with the same signal to noise ratio up to tens of meters. Future plans include experimental verification of these predictions, shortening of system response time and remote detection demonstrations in the field.

8.6. Summary

Enhanced pulsed laser photodissociation/laser-induced fluorescence method for remote detection of explosives vapor and trace quantities was utilized to detect high explosives. We demonstrated the ability of remote detection from 2.5 m of explosives with extremely low vapor pressures such as RDX, PETN and C4. The current system has a response time of up to 10 s and can be successfully applied for distances of several tens of meters. The enhanced PLP/LIF method is applicable to all nitrogen-based explosives and can be extended to additional hazardous materials.

Acknowledgement

The preliminary research was done in collaboration with Ben-Gurion University and was sponsored by agencies of the United States Government and the Government of Israel under the Counterterrorism Research and Development Memorandum of Agreement.

References

[1] Steinfeld, J.I., and J. Wormhoudt, 1998. Explosive detection: A challenge for Physical Chemistry, Annu. Rev. Phys. Chem., 49, 203–232.

[2] Ledingham, K.W.D., 1995. The use of lasers to detect strategic and environmentally sensitive materials, Phys. Scr., T58, 100–103.

[3] Kolla, P., 1995. Detecting hidden explosives, Anal. Chem., 67, 184A–189A.

[4] Singh, S., and M. Singh, 2003. Explosives detection systems (EDS) for aviation security, Signal Process., 83, 31–55.

[5] Moore, D.S., 2004. Instrumentation for trace detection of high explosives, Rev. Sci. Instrum., 75 (8), 2499–2512.

[6] Vartsky, D., G. Engler, and M.B. Goldberg, 1993. A method for detection of explosives based on nuclear resonance absorption of gamma rays in 14N, Proc. SPIE, 2092, 307–316.

[7] Cabalo, J., and R. Sausa, 2005. Trace detection of explosives with low vapor emissions by laser surface photofragmentation-fragment detection spectroscopy with an improved ionization probe, Appl. Opt., 44 (6), 1084–1091.

[8] Wormhoudt, J., J.H. Shorter, J.B. McManus, P.L. Kebabian, M.S. Zahniser, W.M. Davis, E.R. Cespedes, and C.E. Kolb, 1996. Tunable infrared laser detection of pyrolysis products of explosives in soils, Appl. Opt., 35, 3992–3997.

[9] Effenberger, E.F., and A.G. Mercado, 1998. Explosive vapor detection via mid-infrared laser spectroscopy, in Electro-Optical Technology for Remote Chemical Detection and Identification III, edited by M. Fallahi and Howden, Proc. SPIE, 3383, 104–112.

[10] Lewis, I.R., N.W. Daniel, Jr., and P.R. Griffiths, 1997. Interpretation of Raman spectra of nitro-containing explosive materials. Part I: Group frequency and structural class membership, Appl. Spectrosc., 51, 1854–1867.

[11] Carter, J.C., S.M. Angel, M. Lawrence-Synder, J. Scaffidi, R.E. Whipple, and J.G. Reynolds, 2005. Standoff detection of high explosive materials at 50 meters in ambient light conditions using a small Raman instrument, Appl. Spectrosc., 59 (6), 769–775.

[12] Gilbert, B.D., J. Janni, D. Moss, R.W. Field, J.J. Steinberg, K. Kniepp, Y. Wang, R.R. Dasari, and M.S. Feld, 1995. Spectroscopic detection methods for explosive molecules and their fragmentation products, Proceedings of 5th International Symposium on the Analysis and Detection of Explosives, Washington, DC, edited by Ch.R. Midkiff, Bureau of Alcohol, Tobacco and Firearms.

[13] DeLucia, F.C., Jr., R.S. Harmon, K.L. McNesby, R.J. Winkel, Jr., and A.W. Miziolek, 2003. Laser induced breakdown spectroscopy analysis of energetic materials, Appl. Opt., 42 (30), 6148–6152.

[14] Shu, J., I. Bar, and S. Rosenwaks, 2000. The use of rovibrationally excited NO photofragments as trace nitrocompounds indicators, Appl. Phys. B, 70, 621–625.

[15] Arusi-Parpar, T., D. Heflinger, and R. Lavi, 2001. Photodissociation followed by laser-induced fluorescence at atmospheric pressure and 24 °C: A unique scheme for remote detection of explosives, Appl. Opt., 40 (36), 6677–6681.

[16] Heflinger, D., T. Arusi-Parpar, Y. Ron, and R. Lavi, 2002. Application of a unique scheme for remote detection of explosives, Opt. Commun., 204, 327–331.

[17] Wu, D., J.P. Singh, F.Y. Yueh, and D.L. Monts, 1996. 2,4,6-Trinitrotoluene detection by laser-photofragmentation-laser-induced fluorescence, Appl. Opt., 35, 3998–4003.

[18] Bordreaux, G.M., T.S. Miller, A.J. Kunefke, J.P. Singh, F.Y. Yueh, and D.L. Monts, 1999. Development of a photofragmentation laser-induced-fluorescence laser sensor for detection of 2,4,6-trinitrotoluene in soil and groundwater, Appl. Opt., 38, 1411–1417.

[19] Pastel, R.L., and R.C. Sausa, 2000. Spectral differentiation of trace concentrations of NO_2 from NO by laser photofragmentation with fragment ionization at 266 and 452 nm: quantitative analysis of NO–NO_2 mixtures, Appl. Opt., 39, 2487–2495.

[20] Pastel, R.L., and R.C. Sausa, 1996. Detection of NO and NO_2 by (2+2) resonance-enhanced multiphoton ionization and photoacoustic spectroscopy near 454 nm, Appl. Opt., 35, 4046–4052.

[21] Swayambunathan, V., G. Singh, and R.C. Sausa, 1999. Laser photofragmentation–fragment detection and pyrolysis-laser-induced fluorescence studies on energetic materials, Appl. Opt., 38, 6447–6454.

[22] Davies, J.P., L.G. Blackwood, S.G. Davis, L.D. Goodrich, and R.A. Larson, 1993. Design and calibration of pulsed vapor generators for TNT RDX and PETN, in Advances in Analysis and Detection of Explosives, edited by J. Yinon, Kluver Academic Publishers, pp. 513–532.

[23] Griffy, T.A., 1993. A model of explosive vapor concentration II, in Advances in Analysis and Detection of Explosives, edited by J. Yinon, Kluver Academic Publishers, pp. 503–511.

9. GAS ANALYTICAL SYSTEM OF EXPLOSIVE DETECTION FOR THE MANUAL CHECK OF OBJECTS AND USE IN WALK-THROUGH PORTALS

V.M. Gruznov[1] and G.V. Sakovich[2]

[1]*The Design & Technological Institute of Instrument Engineering for Geophysics and Ecology, The Siberian Branch of RAS, Siberia*

[2]*Institute for Problems of Chemical & Energetic Technologies, SB RAS, Siberia*

With a design to decrease detection limit, enhance sensitivity and expand a field of application it is suggested to develop a gas analytical detector with air as a carrier-gas, remote vortex sampling and sample injection with two-stage concentration. The two-stage concentration makes it possible.

- to preset parameters of the concentrator of the first stage (preconcentrator) regardless of the requirements on sample injection into an analyzer, what gives a possibility to increase the flow rate of air pumped through the concentrator, and thus to increase the amount of a compound in a sample without additional expenditures of energy;
- to increase selectivity of analysis owing to the characteristics of the concentrator of the second stage;
- to solve the problem of sample injection from a large size concentrator into a device.

As a base model at the first stage it is suggested to develop a gas chromographic detector with two-stage concentration, an independent remote vortex sampler for manual check of the objects. Detection limit is 10^{-15} g/cm^3.

The next stage is to develop a detector with automated sampling and sample injection through the use of the stationary concentrator of the first stage. Detection time is 10–15 s, detection limit is 10^{-15} g/cm^3. This detector can be used in walk-through portals and systems for monitoring air in restricted areas (metro, railway stations etc).

The advantage of this project is the use of a two-stage automatic concentration in the analytical system itself based on a gas chromatograph. This enhances the sensitivity of the analytical system by an order of magnitude, on the one hand, and simplifies sample injection from the concentrator into the device, on the other hand.

H. Schubert and A. Rimski-Korsakov (eds.), Stand-off Detection of Suicide Bombers and Mobile Subjects, 69–69.

10. THREAT DETECTION DC TO DAYLIGHT

Lyle Malotky, Ph. D.
Transportation Security Administration
Department of Homeland Security, USA

The treat posed to civilized societies by suicide bombers is severe and growing. The need exists to be able to detect these devices at a distance in order to prevent the terrorist from reaching his target. Automated detection is the goal as the number of people at potential target locations to be screened is in the millions per day. The suicide bomb is attractive from the terrorist's perspective because it is lethal, mobile and adaptable to a changing target environment. We have seen instances of human carried suicide bombs being successfully used to kill hundreds of innocent civilians in Israel, Russia, India, England, ...

The development of tools that will allow the standoff detection of these devices that are designed to be both deadly and invisible is a challenge. The probe used must not only be able to penetrate the clothing of the terrorist but detect the unique properties of the explosive. The detection challenge is

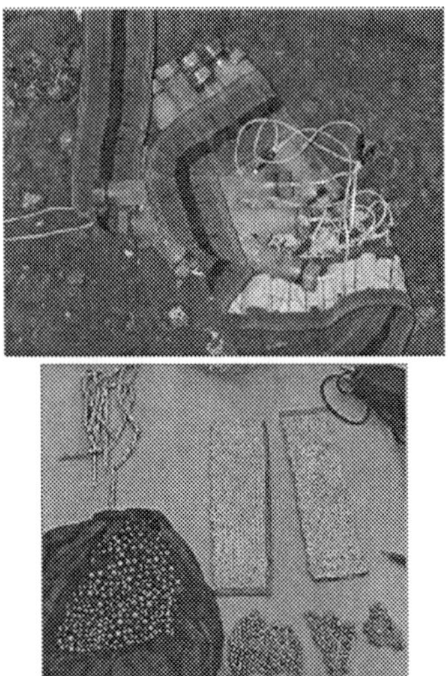

Figure 1. Background on threat

H. Schubert and A. Rimski-Korsakov (eds.), Stand-off Detection of Suicide Bombers and Mobile Subjects, 71–75.

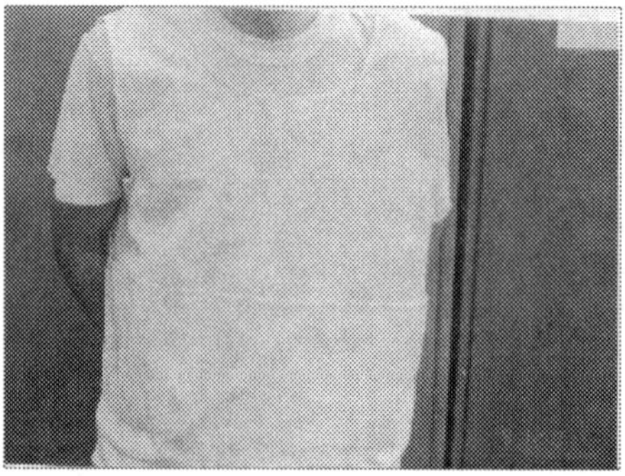

Figure 2. Understanding the threat

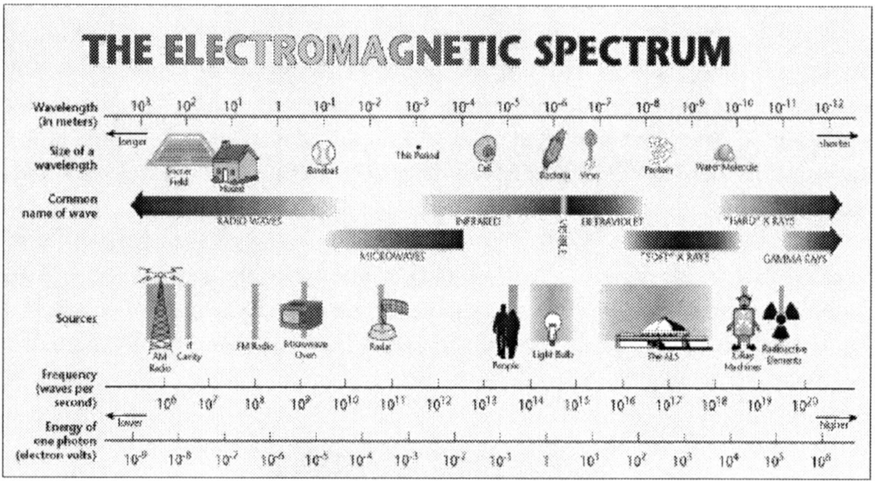

Figure 3. What can we use to see the bomb

Figure 4. Quadrupole resonance portal

Figure 5. Passive MMW imager

complicated by the number of explosives that could be used and their diverse chemical compositions and physical properties. Differentiation of the explosive and other bomb components from innocent objects or medical devices that people may have concealed on their person depends on the ability to sense at a distance unique physical or chemical properties of the explosives.

If we examine the electromagnetic spectra we can identify areas that offer the potential of sensing and/or identifying concealed anomalies at a distance. In the AM radio band, selected explosives can be detected by their quadrupole resonance. This is not a standoff technique. The subject must stand in something like a phone booth for a few seconds, for the automated detection of

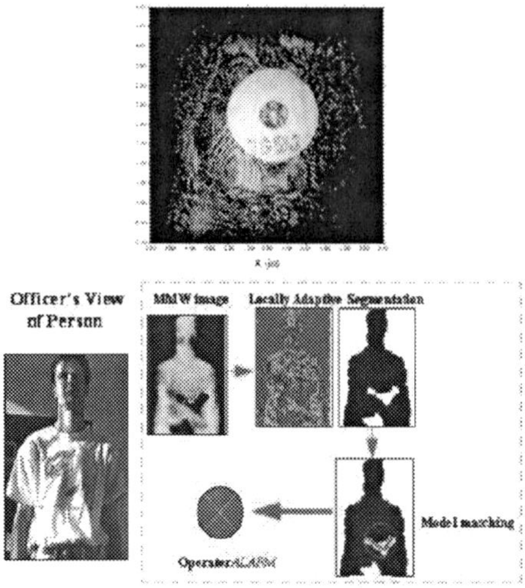

Figure 6. Active millimeter wave

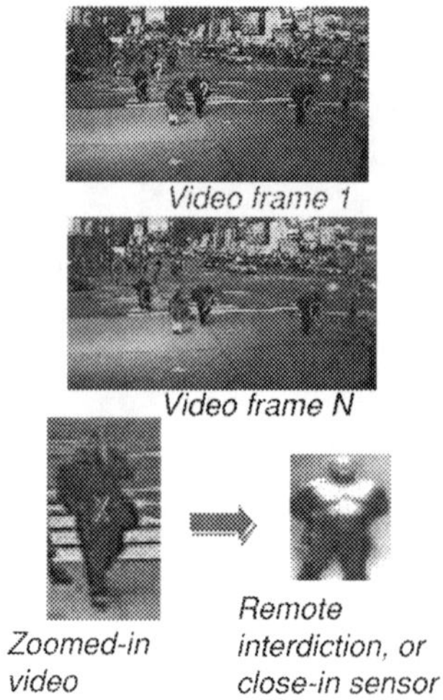

Figure 7. Active millimeter wave with video tracking

small explosive threats concealed on a person to take place. As we move to longer wavelengths in the millimeter region (94 GHz) we can have an imaging system that can penetrate clothing and operate in either the active or passive modes. Sources and detectors in this area are available but the long wavelength leads to large optics. Some vendors claim automated detection of anomalies.

As we move to higher frequencies, in the THz region (300–3000 GHz), the resolution of objects improves (using either active or passive interrogation) as the wavelength gets shorter. The potential for THz spectrometry exists which would allow the extraction of specific spectral features characteristic to the explosives, allowing their detection and differentiation from other common objects. Challenges exist in the THz area with a lack of low cost sources and detectors as well as radiation absorption by water vapour, in some parts of the spectra, limiting range or perhaps even use at some frequencies.

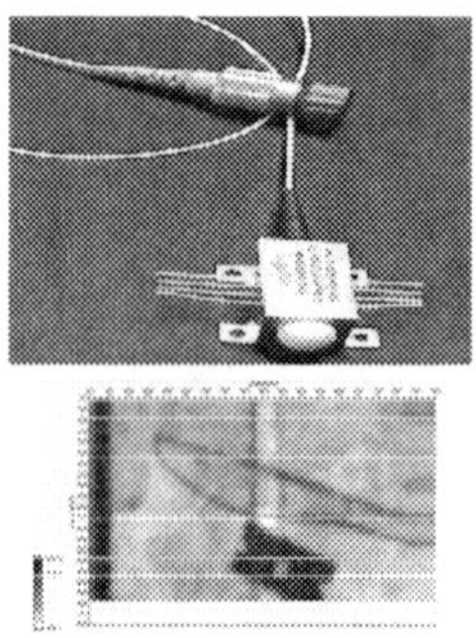

Figure 8. Terahertz imaging/spectroscopy

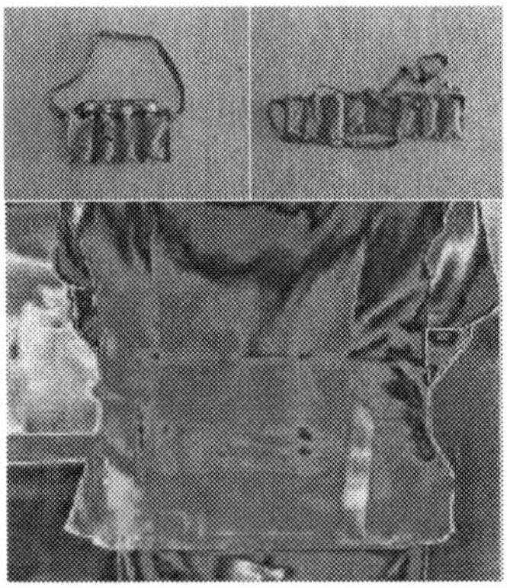

Figure 9. Thermal imager

Moving to still shorter wavelength in the infrared where the body is a strong emitter, imaging is possible with low cost cameras. Penetration of clothing becomes a problem at these shorter wavelengths. In the visible part of the spectra, clothing is nominally opaque however the potential use of active interrogation, employing either remote Raman scattering or laser-induced breakdown spectroscopy, is possible but may require sources more intense than may be eye safe. Visible light may also be used to look for anomalous gate, behaviour, or drape or movement of the clothing, as well as for automatic tracking of a suspect identified by one of the other standoff technologies.

Potential exploitable areas of the electromagnetic spectra have been identified for the detection of bombs concealed on persons. What is needed are better sources and detectors covering the entire range as well as automated pattern recognition and tracking to identify and follow a suspect. Finally for electromagnetic systems to be operationally viable, not only do they need to be effective and reliable but because of the thousands of potential soft targets of a suicide bomber in any city, the entire system must be low cost.

11. STAND-OFF DETECTION OF CONCEALED IMPROVISED EXPLOSIVE DEVICES (IEDS)

Rory Doyle[1] and John McNaboe
Smiths Detection—Ireland, U4, Westpoint Business Park Ballincollig, Co. Cork, Ireland

Abstract. In the millimetre-wave region of the spectrum, many lightweight materials such as clothing fabrics appear as transparent. This property opens up applications in the security area for scanning of personnel for concealed threat objects or contraband. The system described here is a mechanically scanned millimetre-wave imaging camera that can be interchangeably used in an active or passive mode and provides rapidly refreshed, raster imagery with high efficiency across all millimetre-wave frequencies. The proposed use of the imager is in an active configuration to observe a checkpoint or crowd scene at stand-off distances of up to 50 m. Target objects to be detected are the hidden metallic components associated with improvised explosive device (IED) constructions. A typical IED consists of several explosive filled pipes strapped to the body or clusters of small metallic objects embedded in explosives. Trials have yielded positive results by showing the presence of concealed metallic objects on people at distances of 25 m. Objects detected have included simulated IED constructions such as metal pipes and clusters of nuts and bolts. These tests have been conducted using both continuous wave (CW) and frequency modulated CW (FMCW) front-end configurations. The use of FMCW radar introduces the possibility of gathering range as well as intensity information from the scene. This additional range information opens up significant possibilities in pinpointing people carrying threat objects particularly in crowd situations where two-dimensional intensity patterns may not provide sufficient information to identify a threat location.

Key words: millimetre-wave, surveillance, stand-off, radar, active, scanning

11.1. Introduction

In an ideal scenario, imaging a scene using millimetre-wave frequencies would be done using a fully populated focal plane array. However, using currently

[1] E-mail: rory.doyle@smithsdetection.com

H. Schubert and A. Rimski-Korsakov (eds.), Stand-off Detection of Suicide Bombers and Mobile Subjects, 77–87.

available technology the cost and complexity of implementing a full array means only a limited number of applications can be addressed with this approach. An alternative option is to use a single or sub-array of receivers and to scan the scene using an electronic or mechanically driven method. Using a mechanically driven optical scanning method allows for a rapid capture of a given field of view in an efficient and easily interpreted user format.

To exploit the potential of millimetre-wave imaging, a suitable receiving system must be employed to detect and process the appropriate signals in the 100 GHz region of the spectrum. To date, use of sensors for detecting signals at these frequencies has been hindered by cost and performance drawbacks that have limited the effective use of millimetre-wave imaging. Recently available monolithic microwave integrated circuits (MMICs) operating up to 100 GHz combined with scanning systems that focus scene information onto a sub-array of receivers has opened up the possibility of commercially affordable and practically deployable millimetre-wave imagers.

The system described here is focus adjustable to image close-in for portal screening application or at long distances for use in surveillance situations. The emphasis in this paper is on long distance imagery for detection of threat materials on personnel in stand-off surveillance use.

11.2. Scanning System

There are several key performance targets that a millimetre-wave imaging system must achieve to be suitable for a security surveillance application. In order to achieve a suitable spatial resolution, an appropriate optical aperture must be used. The image refresh rate must be made as rapid as possible through the use of a sufficiently fast line or frame scan while maintaining a suitable integration time per pixel. The quality of the optical elements must be maintained to minimise loss and scatter in the optical path. With these constraints in mind, the overall system needs to be compact so that it can be practically deployed in a security or surveillance style application. The system that is described here was developed for use in both portal screening and long distance surveillance applications. The basic parameters are described below.

- *Principle of operation:* Rotating reflectors with an off-axis tilt that focus a line scan of the scene onto the millimetre-wave receiver(s).
- *Optical elements:* mm-wave mirrors ~100% energy reflectance.
- *Optical aperture:* Concave focusing reflector—aperture typical = 600 mm.
- *Field of view:* Vertical: $\pm 12^\circ$ (mechanically variable). Horizontal: 0–20° (dependant on receiver array configuration).

- *Angular resolution:* 0.3° at 94 GHz.
- *Refresh rate:* Up to 20 frames/second.

Rotating mirrors are used to implement the scene scan with an optical path that is purely reflective. By combining tilted plane and concave mirrors whose rotations are equal and opposite a line scan of the scene is produced—one upward and one downward scan for each complete rotation of the mirrors. A third mirror can used to fold the optical path in order to achieve a compact form factor for the imager design.

A 600 mm aperture was chosen based on the desired spatial resolution and the practical limit of rotating the offset mirror at high speeds. This optical aperture fixes the achievable resolution and also determines the overall physical size of the system.

The scanning method (described in Lettington et al., 2004) has a number of key benefits that are listed below. These result in a high quality millimetre-wave image being captured.

- *Reflective optics:* The imager uses only reflective optics in directing and focusing the incident radiation. This ensures that almost the entire signal energy reaches the millimetre-wave sensor—thus maximising the system efficiency and signal to noise ratio for the processing electronics. The fact that there are no frequency dependent optical elements in the signal path means that the camera is independent of the frequency being detected and can be used over the entire millimetre-wave spectrum.
- *High rotation/refresh rates:* The line scan is generated by rotating two circular reflectors at constant speed—each rotating in one direction only. This configuration allows for a high rate of revolution of the mirrors and consequently a fast refresh rate in the image. Based on an array of 16 receivers a 10 Hz refresh rate is achieved.
- *Raster scan:* The rotation of the mirrors generates line scans of the scene. By focusing the line scans onto a parallel array of receivers, a raster image frame is directly generated by the system. The uniformity of the raster output enables image processing to be performed directly on the output.
- Active and passive operation: Because there is no manipulation of signal in the optical path the camera can be configured as either a passive receive only unit or as an active transceiver system.
- *Low Cost:* A low sensor count and simple mechanics combine to provide a physically compact millimetre-wave camera with high performance and low cost of manufacture.

The scanning system is illustrated in Figure 1 where the reflector configuration and motion are shown.

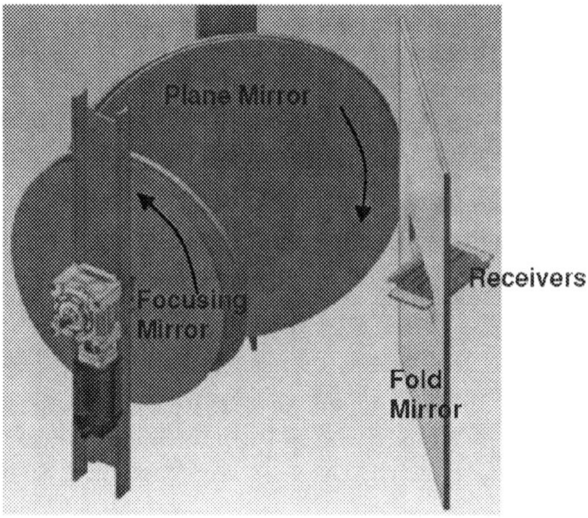

Figure 1. Scanning mechanism showing the configuration and relative motion of the reflectors

11.3. Realisation of the Imaging System

The imaging system has been implemented for both close proximity portal imaging and long range surveillance applications. In the case of the portal imaging the system was configured as a passive imager for indoor use. To achieve a suitable performance in this setting, a thermally generated illumination scheme was included in the design to provide contrast. This system was also used in the development of an active stand-off application where the passive receivers were combined with millimetre-wave sources or replaced with an active FMCW transceiver. In this mode the illumination components associated with the passive system were removed from the design.

A reduced scale system based on an FMCW transceiver was also constructed and is more representative of the type of system that could be used as an easily transported and field deployable unit. The compact design does however have a poorer resolution due to its smaller optical aperture. These systems have been trialed under demonstration conditions to determine the feasibility and performance of a stand-off concealed weapons detection system. Figure 2 shows both the portal and reduced scale imager constructions.

11.4. Stand-Off IED Detection

To image at stand-off distances, the scene needs a greater level of illumination than can be provided through the contrast of cold sky or blackbody/thermal generation. Active sources of millimetre-wave energy are needed to identify targets of interest at distances of up to 50 m. The application areas for this type

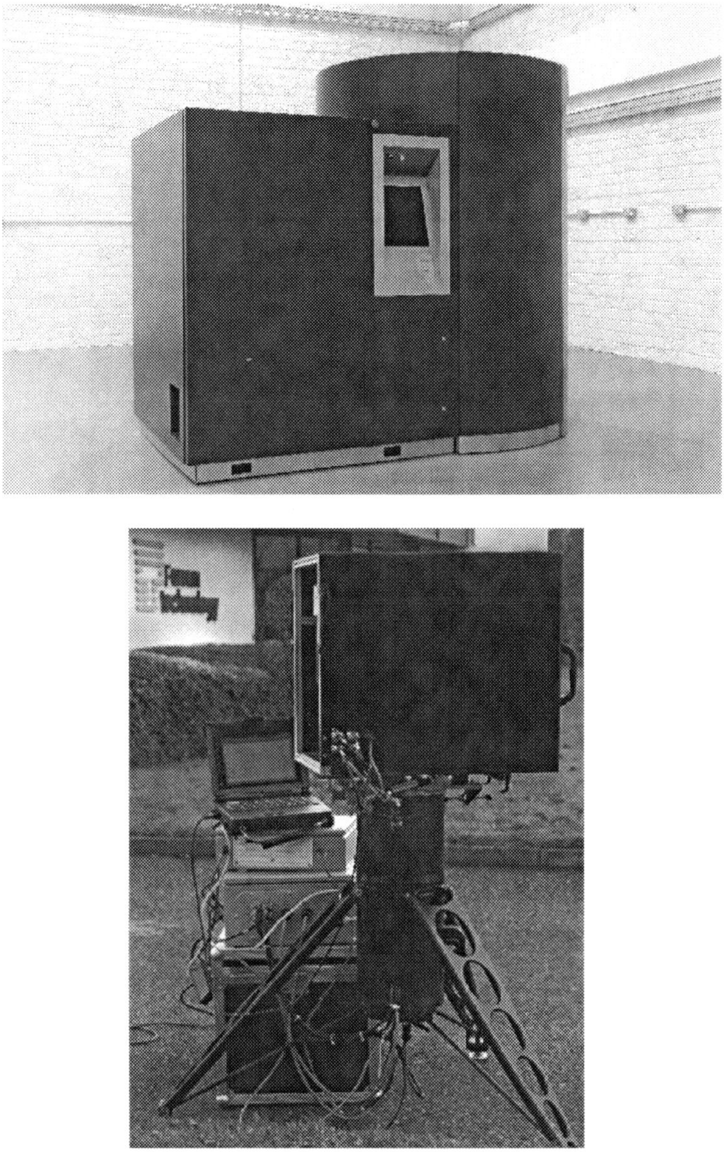

Figure 2. Imager realisations for indoor portal imaging (above) and stand-off surveillance (below)

of imaging are the identification of individuals in a crowd or at a checkpoint who are carrying potential threat objects that are concealed inside clothing, baggage etc. A typical threat in this category is a large metallic weapon or a bomb consisting of metallic pipes and shrapnel. In this situation the monitoring camera and personnel should remain at a safe distance during observation so

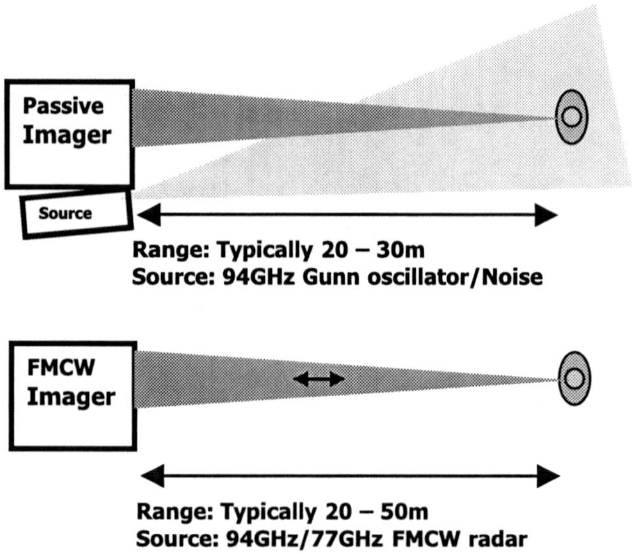

Figure 3. Active imager formats using separate (broadcast) and integrated (focussed) sources of scene illumination

that they are sufficiently out of range of any detected weapon or explosive device.

The options that were used for millimetre-wave imaging were a discrete source that provided general scene illumination or an integrated FMCW transceiver that used the scanner optics to focus the illumination onto the observed scene location. The two alternatives are illustrated in Figure 3.

Two sources were used for the general illumination case based on a coherent Gunn oscillator tuned to 94 GHz centre frequency and a wideband noise source. The noise source succeeded in illuminating close proximity scene objects. However, for the longer distances of interest it did not have sufficient power density to produce the required return signal.

In the second trial, a general broadcast illumination of the scene was set up using a 94 GHz tuned Gunn oscillator. The working distance was set to be 20 m from the imager. Targets used were simulated bomb components—aerosol canisters to simulate pipe bomb constructions and clusters of nuts and bolts to simulate shrapnel. Control of the source power level was a critical factor as the scene was easily saturated by relatively low emission levels. The final setting to achieve discrimination of the target was a very low power—close to 0 dB m. Images were taken of control conditions where a person carrying no metal objects stood in the scene and then of the same scenario except with the person concealing various metallic items. Figures 4 and 5

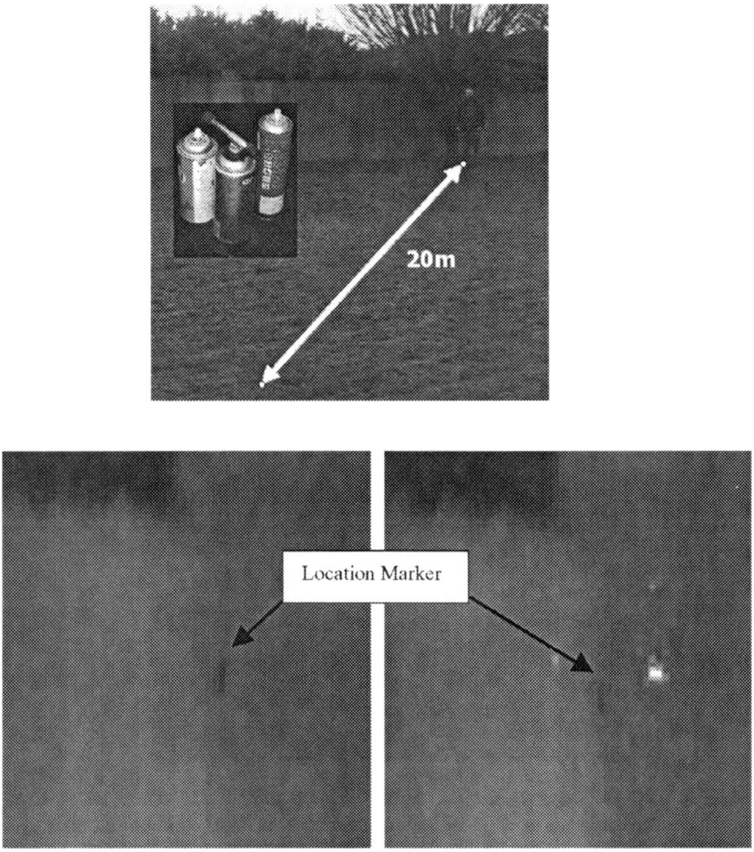

Figure 4. Example of active imaging for stand-off surveillance applications. The visual scene is shown above with an inset of the aerosol cans used to simulate pipe bomb materials. The scene imaged with passive radiometry (lower left) shows no evidence of the person or the concealed metallic objects. Once the Gunn oscillator is energised (lower right), the concealed metallic objects are detected on the person as the bright return in the image

show the results of these trials where the ability to detect personnel carrying suspect items is demonstrated.

A 77 GHz FMCW radar front-end was also installed in the imager where the system optics was used in the transmission and focusing of the outgoing illumination. Using the FMCW radar it is possible to read range as well as intensity information from the scene. This availability of range information opens up significant possibilities in pinpointing people carrying threat objects particularly in crowd situations where two-dimensional intensity patterns may not provide sufficient information to identify a threat location.

Figure 5. An example showing the capability of the camera to identify the individual on the right who is carrying an assortment of metallic objects (nuts, bolts etc) concealed under clothing. The bright echo corresponds to the person carrying the metal whereas the person on the left does not show up in the image. The dark areas in the millimetre-wave image are the reflection of sky off of the metal strips that are used as location references

The ranging capability of FMCW meant that it was possible to get a good accuracy on locations of concealed threat objects in the scene. Limitations on the VCO quality placed an outer boundary of approximately 20 m on the working distance of the radar. However, once this is corrected it is expected that distances of over twice this range will be possible. The field of view of the imager was restricted to give a view of approximately 4 m square at the 20 m distance. Actual scanning time for such a window would be in the order of 2–4 s in an optimised system—redundancy in the current configuration meant a large part of the sweep was not used for actual imaging.

The results confirmed the expected capability of the system to distinguish between people carrying metallic objects and those without such material. An example of the radar output is shown in Figure 6 where two individuals are screened and the one carrying a number of concealed aerosol canisters is

Figure 6. FMCW imaging to identify people carrying threat objects in this example. The top photograph shows the visual scene. The person on the right hand side is concealing several aerosol canisters (similar to those in the Figure 4 inset). The centre image is of the normalised output from the radar where the dark background indicates areas of low level response and the lighter shadings show areas of return where metallic objects may be present. Removing the low level returns and overlaying the peak responses onto the visual picture gives an indication of a potential user interface (lower picture)

clearly identified. The corresponding range information was also correlated to the location of the subjects.

11.5. Applications

Stand-off detection of IEDs using millimetre wave imaging has a direct application in areas such as checkpoint and public area surveillance where suspected threat objects may be carried by people. Early identification of IEDs affords the user a level of safety once the detection has been made. The typical situation would be the surveillance of a checkpoint or gateway where people are passing through a controlled area. In this case a continuous monitoring of the checkpoint can be set up by making the field of view equivalent to that through which the people are passing. An alternative situation is where a more dispersed group of people need to be monitored such as in an open public area or on a street. In order to monitor such a scene with a real time refresh rate it may be more practical to sweep a narrow window field of view across the scene. Combining a swept monitoring window with a visual video view of the scene would give a very effective user interface. This is illustrated in Figure 7.

While both the broadcast illumination using the Gunn oscillator and the focussed method using the FMCW have been shown to work for these

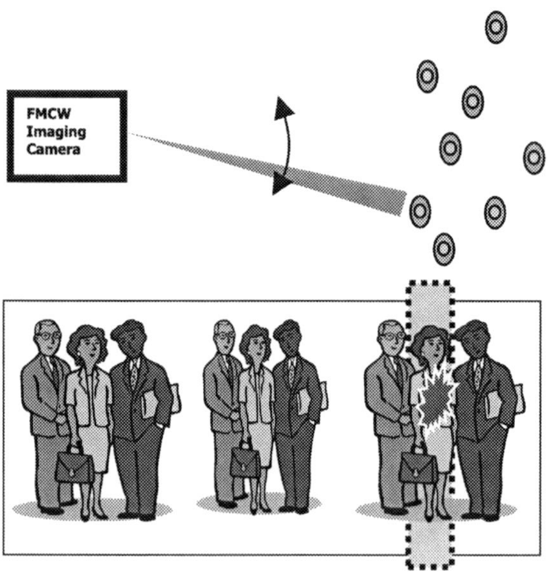

Figure 7. A representation of the use of a surveillance imager is shown here. The scanned window is swept across a crowded public area. The millimetre-wave output is fused with its corresponding video image of the scene. By showing only the high level returns from the scene as an overlay on the video display a very easily interpreted user interface can be implemented

applications, the addition of range information provided by the radar gives it a key advantage. In particular the public area application where identification of an individual in a randomly moving crowd situation would need range data to accurately locate a person carrying a suspect item.

11.6. Conclusion

Having demonstrated the ability to detect suspect materials carried on people at a distance, the evolution of the system to a field deployable imager is an immediate objective. Depending on the intended usage, the imaging camera can be configured to suit the applications described in the previous section. The fusing of millimetre-wave imager with a visual picture of the scene has already been successfully demonstrated for still images. The extension of the camera to real time imaging and the fusion of this output with a standard video feed will lead to a readily usable product.

Trials to date have worked on surveillance distances of up the 30 m. The extension of this distance to 50 m will be demonstrated once upgrades to both the front-end and camera performance have been implemented. To date the camera has been used for surveillance applications at both 77 and 94 GHz. Further work with different frequency ranges is expected to yield a better characterisation on the detection and discrimination of different target types.

The capability to identify concealed threat objects in stand-off surveillance situations has been demonstrated. Different options for illumination of distant targets have been evaluated and both broadcast CW and focussed FMCW approaches have been shown to be valid. The successful identification of concealed metallic targets at ranges of greater than 20 m has been carried out and the extension to distances up to 50 m is planned. User interfaces that will allow the easy interpretation of millimetre-wave imagery have been proposed and implemented in a still image format.

The application of millimetre-wave imaging to stand-off security screening offers an effective means of monitoring personnel at distances that can maintain safety for the observer and give adequate warning of potentially dangerous situations.

References

Lettington, A.H., D. Dunn, N.E. Alexander, A. Wabby, B.N. Lyons, R. Doyle, J. Walshe, M. Attia, and I. Blankson, 2004. Design and development of a high performance passive mm-wave imager for aeronautical applications, Proc. SPIE, Orlando, Vol. 5410, pp. 210–218.

12. STAND-OFF DETECTION OF SURFACE CONTAMINATIONS WITH EXPLOSIVES RESIDUES USING LASER-SPECTROSCOPIC METHODS

Reinhard Noll and Cord Fricke-Begemann
Department of Laser Measurement and Testing Technology, Fraunhofer Institute for Laser Technology, Steinbachstr. 15, 52074 Aachen, Germany

Abstract. Laser-spectroscopic methods have a high potential for the remote detection of surface contaminations with residues of explosives, since they can operate contactless, fast and sensitive. Laser-induced breakdown spectroscopy and Raman spectroscopy are two versatile methods which can be employed to identify a wide variety of materials and for which stand-off operation has been demonstrated. Since both methods provide complementary spectroscopic information while using a similar instrumentation they can be combined to further increase the selectivity and sensitivity as necessary for security applications.

Key words: laser spectroscopy, remote LIBS, laser-induced breakdown spectroscopy, Raman spectroscopy, LIBS-Raman combination, explosives detection, stand-off detection

12.1. Introduction

The handling and manufacturing of explosive materials often results in the contamination of surfaces with residues of the explosive which cannot be removed easily. Contaminated hands of a person who has handled explosives can leave traces, for example, at the door handles of a car which has been loaded with such material, on certain parts of the luggage, cloths and documents the person has touched, or on the wrapping of improvised bombs that may have been placed in public areas. Such residues can be used as an approach for the detection of explosive devices and the prevention of planned terroristic attacks.

Optical analytical methods are predestined for stand-off detection applications since they operate without contact, fast, without sample preparation,

[1] E-mail: reinhard.noll@ilt.fraunhofer.de

H. Schubert and A. Rimski-Korsakov (eds.), Stand-off Detection of Suicide Bombers and Mobile Subjects, 89–99.

and with little or no sample destruction. Furthermore, they have matured in a variety of applications over the past decades and instrumental key components are commercially available.

Many common optical detection methods are based on the specific absorption by explosives in the ultraviolet or mid-infrared spectral range and can be employed for the detection of these substances in the vapor phase. However, many explosives have a very low vapor pressure and the stand-off detection in outdoor ambient air conditions is problematic. On the contrary, for the detection of surface contaminations the promising two methods are discussed below, which use a near-infrared laser for excitation and analyze spectrally the light emitted from the substance under investigation.

12.2. Laser-Induced Breakdown Spectroscopy (LIBS)

Laser-induced breakdown spectroscopy (LIBS) is a method which employs a short laser pulse to vaporize a small amount of material from the surface to be inspected, excite the vapor, and detect the emitted light within microseconds. From the analysis of the optical emission it is possible to derive the chemical composition of the ablated material. While the emission wavelengths reveal the emitting species, the relative intensities provide information on their concentrations.

The possibility of a simultaneous multi-species detection, which can include up to 92 elements, yields the high versatility of this spectroscopic method. At the same time, a single measurement takes less than 50 μs and can be repeated at frequencies up to 1 kHz, thus allowing measurements of moving objects. LIBS analysis can reach detection limits of a few ppm. Considering a total mass ablation in the order of 1 μg, this corresponds to a detected mass of the substance to be determined in the sub-nanogram range.

We have demonstrated and implemented systems using LIBS analysis for a broad range of industrial applications (Noll et al., 2001), as e.g.:

(a) The thickness of surface coatings on metal sheets ranging from 3 to 11 μm has been measured with sub-micron accuracy (Balzer et al., 2005). The developed method can be applied to sheet steel moving with speeds of up to 2 m/s.
(b) The composition of small particles was analyzed in the top gas of a blast furnance (Brysch et al., 2002). Onsite tests at three European blast furnaces demonstrated the capability of LIBS to be deployed under harsh industrial environments.
(c) LIBS identification is used to prevent mix-ups of 30 different materials. More than 1.5 million pipe-fittings have been inspected by a continuously operating facility (Mönch et al., 2000; Noll et al., 2005; Sturm et al., 2006).

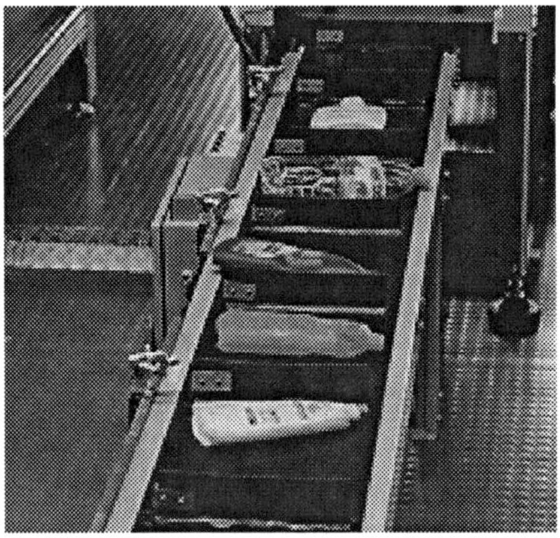

Figure 1. Polymer identification for material recycling

(d) For the material recycling of technical polymers, material identification was demonstrated on basis of the organic compounds (Sattmann et al., 1998) or the concentration of flame retardants (Stepputat and Noll, 2003). In both cases moving parts on a belt conveyor were measured applying a fast autofocussing unit to adjust the focus of the analyzing laser beam onto the surface of the specimen.

In the case of polymer identification, plastic waste bottles were analyzed on a moving conveyer belt, see Figure 1. Using LIBS, the relative concentration of carbon and hydrogen line emissions was measured and compared to the stoichiometric ratio of technical polymers (PE, PP, PVC, PET), see Figure 2. Identifications accuracies of 90–100% were achieved (Sattmann et al., 1998).

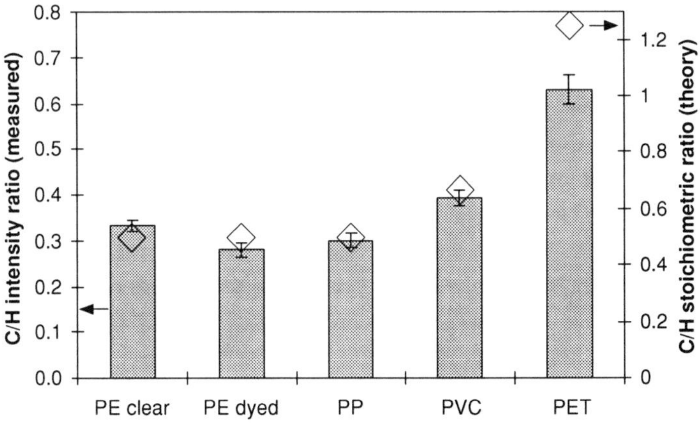

Figure 2. C/H ratios in polymers (bars to left axis: LIBS measurement; diamonds to right axis: theoretical values)

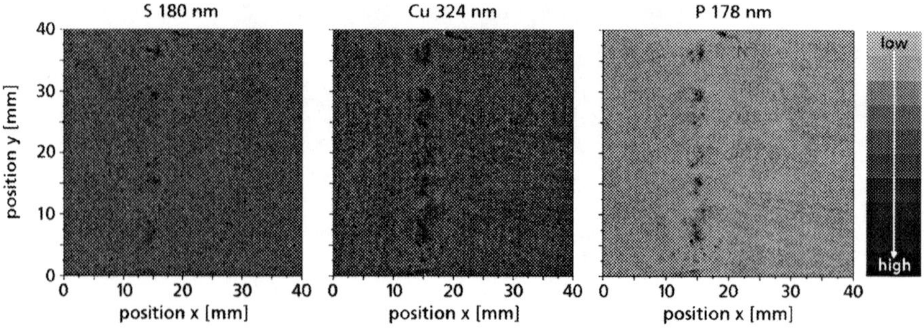

Figure 3. Element maps obtained with a scanning LIBS instrument show localized high concentrations of sulfur, copper, and phosphorus over a 4 × 4 cm^2 area of a steel sample

The small diameter of the focused laser beam together with the high measurement repetition rate allow for a fast scan of the surface under investigation to provide maps of the composition (Bette and Noll, 2004). The necessary high repetition rates have become possible with the laser technology and signal processing electronics developed at the Fraunhofer Institute for Laser Technology. Surface scans with 1000 individual measurements per second and a spatial resolution of 20 μm were described by Bette et al. (2005) and an example of elemental mapping is shown in Figure 3. For security applications this technology yields the possibility to detect localized surface contaminations like fingerprints.

12.2.1. EXPLOSIVES DETECTION

Explosive materials mainly consist of C, H, O, and N, and the quantitative relationship between these elements is the key to determine the type of explosive. For example, stoichiometric nitrogen contents of 13–37% and oxygen contents above 40% are typical characteristics and may even be used to indicate the presence of previously unknown explosives.

Research in the last years has demonstrated that explosives can be identified and discriminated from other materials using LIBS (e.g., DeLucia et al., 2003). Test measurements of substances relevant for explosives detection with a non-calibrated LIBS set-up are presented in Figure 4. The measuring distances were 6 and 25 cm for the emitting and receiving optics, respectively. The laser pulse energy was 5 mJ. Ammonium nitrate (AN), TNT and DNMB are clearly discriminated using the element signals of oxygen, carbon, and hydrogen. While AN contains no carbon, TNT shows the lowest hydrogen emission. The nitrogen content provides another criterion for discrimination but is not considered in this figure. The theoretical stoichiometric relation for the materials used is given in Table 1 for comparison.

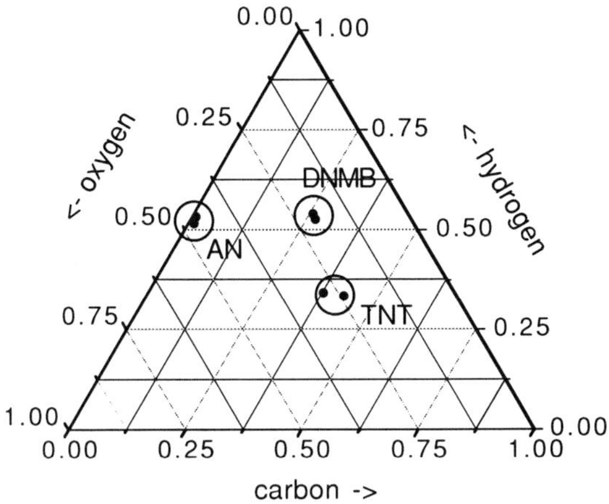

Figure 4. Measured relation of the line intensities of oxygen, carbon and hydrogen in ammonium nitrate (AN), TNT, and DNMB, normalized to the sum of the three components. Two measurements were made for each material

TABLE 1. Relative stoichiometric composition of AN, TNT, and DNMB

Material	Totals formula	C	H	N	O
AN	$H_4N_2O_3$	—	0.44	0.22	0.33
TNT	$C_7H_5N_3O_6$	0.33	0.24	0.14	0.29
DNMB	$C_6H_{12}N_2O_3$	0.26	0.52	0.09	0.13

12.2.2. INFLUENCE OF THE AMBIENT AIR

Together with the material from the surface under investigation, a small volume of the ambient air directly above the surface is excited by the laser pulse and leads to an additional emission. Since the species of interest for the detection of explosives (C, H, O, and N) are also contained in ambient air, measures have to be taken to suppress a possible background. This can be achieved by using multiple laser pulses to generate a region of low gas pressure above the surface. In principle, a first pulse produces a so-called laser generated vacuum while the second pulse generates the emission for the spectroscopic analysis some microseconds later (Sattmann et al., 1995; Noll et al., 2004). This effect is depicted in Figure 5 showing the results of LIBS measurements in an argon atmosphere. Single and double laser pulses were irradiated on a steel target having the same total energy (160 mJ in the case of single pulses and 2 × 80 mJ for the double pulses). The argon line intensity originating from the ambient gas is clearly reduced when double pulses are used. The intensity at standard pressure for double pulses equals that for single pulses at a pressure of about

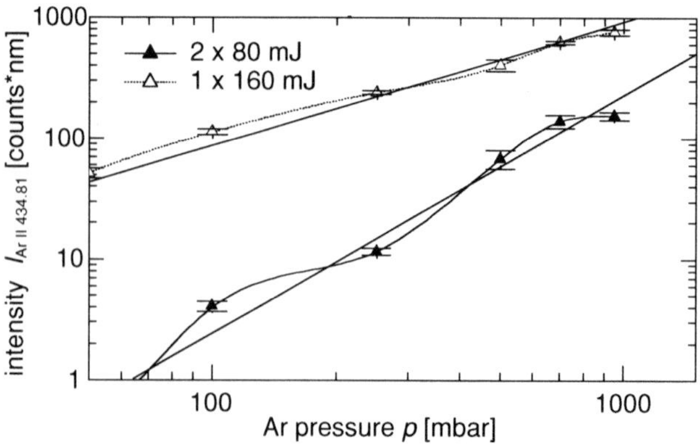

Figure 5. Measured effect of gas pressure on signal intensity from ambient gas (open symbols—single pulses; filled symbols—double pulses)

200 mbar. In addition to this approach, evaluating time-resolved spectra can also help to suppress ambient gas influences and also increase the selectivity of the method since the knowledge of emission decay times may improve the discrimination of different materials (Romano et al., 2004).

12.2.3. STAND-OFF DETECTION

Since LIBS is an analytical technique that operates without contact and uses the same optical path for excitation and observation of the emission, it can easily be used in a stand-off setup. While the typical object distance used so far in most cases for LIBS is in the 10 cm range the good focussability of the laser beam allows for much larger distances. A limitation arises mainly from the received light intensity and the diameter of the optics that can be used in practice. LIBS applications have been demonstrated at distances of tens of meters and several hundred meters have been discussed in the literature (Palanco and Laserna, 2004).

Even the detection of explosives residues over a distance of 30 m has recently been demonstrated (Miziolek and DeLucia, 2005; Lopez-Moreno et al., 2005).

A transportable LIBS instrument for remote material analysis over distances up to 20 m has been developed in cooperation between the Fraunhofer Institute for Laser Technology and the *Laser Analytical Systems and Automation GmbH* (Aachen, Germany) and is commercially available since end of 2005, see Figure 6.

With this system the effect of double pulses has been investigated over a distance of 12 m using a steel sample, see Figure 7. Line emissions of 13

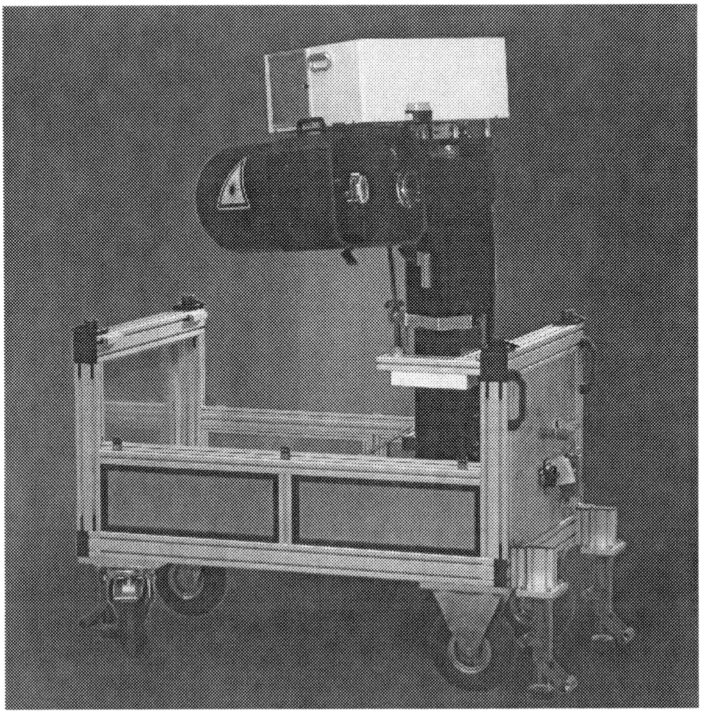

Figure 6. Photo of the transportable LSA TeleLis system. The total height is about 1.8 m

elements were measured simultaneously. Compared to single pulses with the same total energy, the intensities are largely enhanced (up to a factor of 10) and the standard deviation is reduced significantly (note the logarithmic scale of the vertical axis of Figure 7).

To test the system for security applications we have performed measurements over 10 m distance with ammonium nitrate and DNT. Figure 8 shows the single shot spectra in the range of a carbon emission line. While a clear carbon signal was obtained for DNT, there is no signal for ammonium nitrate, i.e., no carbon signal from the ambient air was measured.

12.3. Combination of LIBS with Raman Spectroscopy

Raman spectroscopy is a well-known analytical technique, which has been described in various textbooks. The Raman spectra of explosives and other substances are available as commercial databases and in the open literature. Raman spectroscopy identifies molecules by their inelastic scattering of incident light. The wavelength shift of the laser wavelength is specific for a

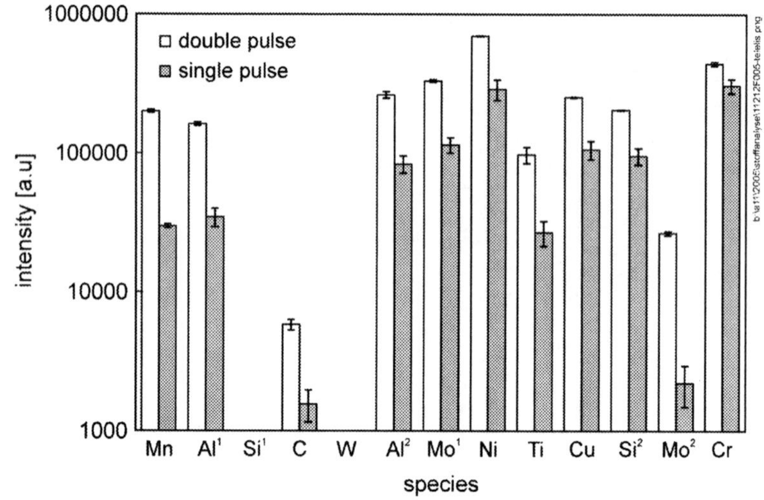

Figure 7. Single and double pulse LIBS signal intensities for different element lines over 12 m distance *with the LSA TeleLis system*

certain molecule. Nitroaromatics, for example, are characterized by the nitrate vibration modes with shifts of about 1350 and 820 cm^{-1}. Recently, the detection of bulk material containing explosives over a distance of 50 m has been demonstrated (Carter et al., 2005). A drawback of Raman spectroscopy is the low backscatter coefficient which reduces the applicability to relatively short distances when used as a stand-alone technique for trace detection.

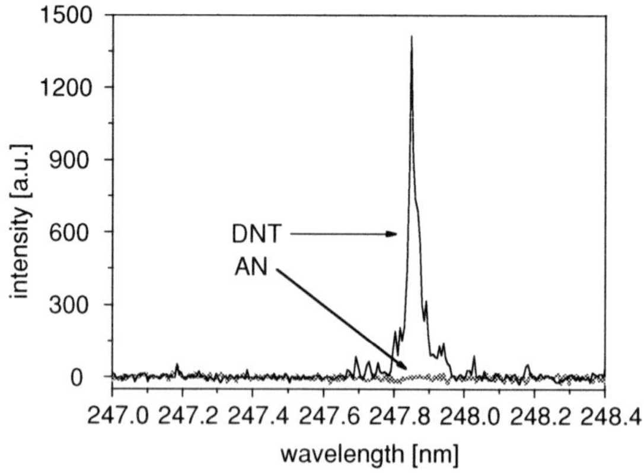

Figure 8. Single shot spectra of the carbon line at 247.856 nm, distance 10 m, of ammonium nitrate and DNT

Raman spectroscopy provides complementary information to LIBS and both techniques can easily be combined in a single instrument since they use similar instrumentation. The combination of LIBS and Raman has been demonstrated for stand-off applications (e.g., Hubble, 2001; Wiens, 2005) and proposed for extraterrestrial explorations (Wiens, 2000). The combination of two complementary analysis techniques can provide a significantly enhanced selectivity for the detection of trace substances.

12.4. Concluding Remarks

Modern security concerns have led to a high demand for stand-off detection methods to prevent assaults with explosive material. Highly developed optical techniques are available to probe surfaces for contaminations with explosives residues. Both laser-induced breakdown and Raman spectroscopy have been used to identify explosives and discriminate them from other materials, and both have been shown to be operable reliably and over distances of tens of meters. Since the two methods provide complementary information, their combination can yield increased selectivity and sensitivity for the detection of explosives. This is necessary for security applications due to the large variety of explosives and background materials, combined with the demand for extremely low false alarm rates and high detection probabilities.

Reliable remote detection of explosives is a challenge for modern analytical technology. Although some optical detection techniques have a high potential this challenge will not be met in the near future by one single detection method alone. However, as stated at the EMSLIBS 2005 security workshop (Schade and Harmon, 2005), *for security applications, only the combination of different laser-spectroscopic and optical detection techniques will give the most efficient impact for engineering the next generation of laser sensors for explosives, IEDs and other hazardous substances.*

References

Balzer, H., M. Hoehne, V. Sturm, and R. Noll, 2005. Online coating thickness measurement and depth profiling of zinc coated sheet steel by laser-induced breakdown spectroscopy, Spectrochim. Acta B, 60, 1172–1178.

Bette, H., and R. Noll, 2004. High-speed laser-induced breakdown spectrometry for scanning microanalysis, J. Phys. D: Appl. Phys. 37, 1281–1288.

Bette, H., R. Noll, G. Müller, H.-W. Jansen, Ç. Nazikkol, and H. Mittelstädt, 2005. High-speed scanning laser-induced breakdown spectroscopy at 1000 Hz with single pulse evaluation for the detection of inclusions in steel, J. Laser Appl. 17, 183–190.

Brysch, A., V. Sturm, R. Noll, H. Denecke-Arnold, H. Brinkmann, K. Mülheims, A. Opfermann, K. Mavrommatis, H.W. Gudenau, and K. Heinänen, 2002. Laser-based elemental analysis of the top gas of a blast furnace, VDI-Bericht, Int. Symposium on Photonics in Measurement, Bericht-Nr. 1694, pp. 117–123.

Carter, J.C., S.M. Angel, M. Lawrence-Snyder, J. Scaffidi, R.E. Whipple, and J.G. Reynolds, 2005. Standoff detection of high explosive materials at 50 meters in ambient light conditions using a small Raman instrument, Appl. Spectrosc., 59 (6), 769–775.

DeLucia, F.C. Jr. R.S. Harmon, K.L. McNesby, R.J. Winkel Jr., and A.W. Miziolek, 2003. Laser-induced breakdown spectroscopy analysis of energetic materials, Appl. Opt., 30, 6148–6152.

Hubble, H.W., M. Ghosh, S.K. Sharma, K.A. Horton, P.G. Lucey, S.M. Angel, and R.C. Wiens, 2002. A combined remote LIBS and Raman spectroscopic study of minerals. Lunar Planet. Sci. XXXIII, 1935.

Lopez-Moreno, C., S. Palanco, J.J. Laserna, F.C. DeLucia, J. Miziolek, A.W. Miziolek, J. Rose, R.A. Walters, and A.I. Whitehouse, 2005. Stand-off detection of explosive residues on solid surfaces with laser-induced breakdown spectroscopy, Presented at the EMSLIBS2005, Aachen, 6–9, September 2005.

Miziolek, A.W., and F.C. DeLucia, 2005. Overview of Laser Induced Breakdown Spectroscopy (LIBS)—An Emerging Versatile Sensor Technology for Counter-IED Applications, Presented at the Counter-IED NATO Meeting, Norfolk, VA, 6–7, April 2005.

Mönch, I., R. Noll, R. Buchholz, and J. Worringer, 2000. Laser identifies steel grades, Stainless Steel World, 4, 25–29.

Noll, R., H. Bette, A. Brysch, F. Hilbk-Kortenbruck, O. Klein, M. Kraushaar, I. Mönch, L. Peter, and V. Sturm, 2001. Laser-induced breakdown spectrometry—from R&D to industrial use: an overview of LIBS applications for process control and quality assurance in steel industry, Spectrochim. Acta B, 56, 637–649.

Noll, R., R. Sattmann, V. Sturm, and S. Winkelmann, 2004. Space- and time-resolved dynamics of plasmas generated by laser double pulses interacting with metallic samples, J. Anal. At. Spectrom. 19, 419–428.

Noll, R., I. Mönch, O. Klein, and A. Lamott, 2005. Concept and performance of inspection machines for industrial use based on LIBS, Spectrochim. Acta B, 60, 1070–1075.

Palanco, S., and J. Laserna, 2004. Remote sensing instrument for solid samples based on open-path atomic emission spectrometry, Rev. Sci. Instrum., 75, 2068–2074.

Romano, C., S. Gräser, K. Faulian, W. Schade, and G. Holl, 2004. Applications of LIBS spectroscopy for remote bulk detection of explosives, edited by H. Schubert, A. Kuznetsov, Detection of bulk explosives, pp. 167–173.

Sattmann, R., V. Sturm, and R. Noll, 1995. Laser-induced breakdown spectroscopy of steel samples using multiple Q-switch Nd:YAG laser pulses, J. Phys. D: Appl. Phys. 28, 2181–2187.

Sattmann, R., I. Mönch, H. Krause, R. Noll, S. Couris, A. Hatziapostolou, A. Mavromanolakis, C. Fotakis, E. Larrauri, and R. Miguel, 1998. Laser-induced breakdown spectroscopy for polymer identification, Appl. Spectrosc. 52, 456–461.

Schade, W., and R. Harmon, 2005. Conclusions of the Security Workshop at the 3rd Euro-Mediterranean Symposium on Laser-Induced Breakdown Spectroscopy, held at Fraunhofer ILT, Aachen, Germany, September 6, 2005.

Stepputat, M., and R. Noll, 2003. On-line detection of heavy metals and brominated flame retardants in technical polymers with laser-induced breakdown spectrometry, Appl. Opt. 42, 6210–6220.

Sturm, V., A. Brysch, R. Noll, H. Brinkmann, R. Schwalbe, K. Mülheims, P. Luoto, P. Mannila, K. Heinänen, D. Carrascal, L. Sancho, A. Opfermann, K. Mavrommatis, H.W. Gudenau, A. Hatziapostolou, and S. Couris, 2006. Online multi-element analysis of the top gas of a blast furnance by LIBS. Proceedings of the 7th international workshop on Progress in Analytical Chemistry in the Steel and Metal Industries, 16–18 May 2006, J. Angeli (ed.), Verlag Glückauf Gmbh, Essen, Germany, pp. 183–188.

Wiens, R.C., D.A. Cremers, M. Ferris, J.E. Nordholt, J.D. Blacic, P. Lucey, and S.K. Sharma, 2000. Development of a prototype laser-induced breakdown spectroscopy (LIBS) instrument with stand-off Raman capabilities as part of the Mars Instrument Development Program. Lunar Planet. Sci. XXXI, 1468, The Lunar and Planetary Institute, Houston.

Wiens, R.C., S.K. Sharma, J. Thompson, A. Misra, and P.G. Lucey, 2005. Joint analyses by laser induced breakdown spectroscopy and Raman spectroscopy at stand-off distances, Spectrochim. Acta A, 61 (10), 2324–2334.

13. DISTRIBUTED TRACE LEVEL EXPLOSIVES SENSOR NETWORK

Pierre Strömbeck[1]
Portendo AB, Positionen 123, SE-11574 Stockholm, Sweden

13.1. Pervasive Explosives Detection

Preventing terrorist attacks is the number one priority for most homeland security offices today. Explosives are the weapon of choice for most terrorists and the target is any populated area. Explosives detection is an active research field with several commercial offspring's. However, very little is done on the system application side. To be practical and useful in the field, sensors need to be a part of a complete system solution that is adaptable to different user's needs and skills.

A *distributed trace level explosives sensor network* combines advanced trace level explosives sensor technology with a distributed network approach. A wide-spread network of low-cost autonomous sensors have the ability to provide a pervasive explosives detection system with early-warning capability. This paper describes the requirements for a pervasive explosives detection system, both on the sensor technology and the system technology. These requirements, stemming from military expertise of the actual problem, were addressed in the design of the Portendo TDS (Threat Detection System).

The pervasive explosives detection solution can be illustrated by the security of a sports event in a sports arena, see Figure 1. Explosives detectors are placed throughout the arena, at the entrances, among the seats and in other important places. All detectors report continuously and in real-time their status to an operations center. The information gathered provides intelligence for optimized apprehension of the criminals.

The example above illustrates one possible application but a similar system could be just as effective in a number of other situations, such as baggage and passenger screening at airports, train stations, subway systems and important buildings as well as screening of cargo and mail.

However, a pervasive explosives detection system requires a combination of leading-edge explosives detection methods and a novel approach to the system application side.

[1] E-mail: pierre.strombeck@portendo.com

H. Schubert and A. Rimski-Korsakov (eds.), Stand-off Detection of Suicide Bombers and Mobile Subjects, 101–106.

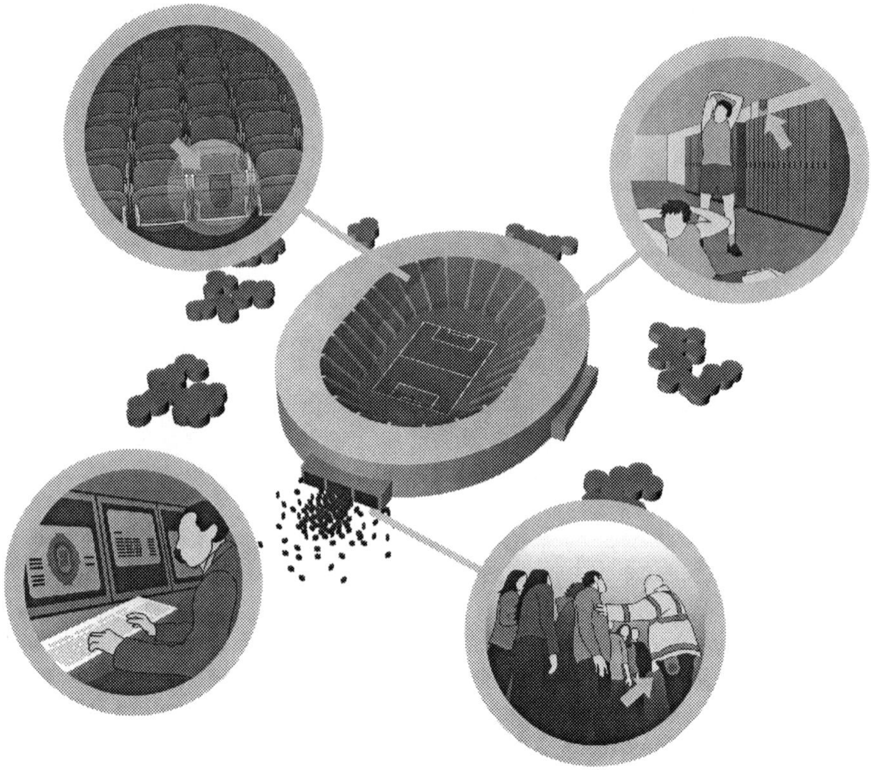

Figure 1. Example of pervasive explosives detection in sports arena

13.2. Sensor Technology

Today, dogs are still superior to commercially available trace level gas detectors. The reason for this is the lack of sensitivity combined with a lack of selectivity in available sensors. It is important to be able to single out different trace gases and unambiguously identify them. Several commercial detectors are based on molecular adsorption onto surfaces. This gives a small change to the resistance which can subsequently be measured. The issue is to develop a surface that only adsorbs the trace gas of interest, but so far this has been impossible. Even if the surface could be 100% selective towards a gas molecule the sensor based on this concept can only detect one type of molecule per sensor.

The concept of applying some type of spectrometry would allow for a selective detection of several molecules at the same time. Selective identification of molecules can be done with infrared spectrometry in the so called "fingerprint region" of light with a 5–16 micron wavelength. In this region the molecular vibration absorption levels are detected and clearly define which molecules are studied. However, the light sources, optics and detectors in this region become large and expensive for other than science or expensive single-device installations.

Raman spectrometry is the solution to smaller devices which can be operated in the visible to near infrared (NIR) regions. In Raman spectroscopy light from a fixed laser wavelength interact in-elastically with gas molecules which scatter light with less energy. The in-elastically scattered light can be detected and shows similar patterns as in the IR fingerprint region. In this way gas molecules can be readily identified.

The main problem using a Raman spectrometer for detection of ultra low concentrations in the gas phase is the low sensitivity of the technique. In normal Raman spectroscopy only 1 out of 10^7 photons are Raman (in-elastically) scattered. However, the Raman signal can be amplified by the use of certain surfaces where surface-enhanced Raman scattering (SERS) occurs. The Raman scattering from a substance (or ion) adsorbed on or even within a few Angstroms of a structured metal surface can be 10^6 times greater than in normal Raman spectroscopy. This surface-enhanced Raman scattering is strongest on silver, but is observable on gold and copper as well. At practical excitation wavelengths, enhancement on other metals is unimportant. Surface-enhanced Raman scattering arises from two mechanisms:

An enhanced electromagnetic field produced at the surface of the metal which occurs when the wavelength of the incident light is close to the plasma wavelength of the metal, conduction electrons in the metal surface are excited into an extended surface electronic excited state called a surface plasmon resonance. Molecules adsorbed or in close proximity to the surface experience an exceptionally large electromagnetic field. Vibrational modes normal to the surface are most strongly enhanced.

The second mode of enhancement is by the formation of a charge-transfer complex between the surface and gas molecule. The electronic transitions of many charge transfer complexes are in the visible region, so that resonance enhancement occurs.

Molecules with lone-pair electrons or pi clouds show the strongest SERS response. The effect was first discovered with pyridine. Other aromatic nitrogen or oxygen containing compounds, such as aromatic amines or phenols, are strongly SERS active.

The intensity of the surface plasmon resonance is dependent on many factors including the wavelength of the incident light and the morphology of the metal surface. The wavelength should match the plasma wavelength of the metal. This is about 382 nm for a 5 μm silver particle, but can be as high as 600 nm for larger ellipsoidal silver particles. The plasma wavelength is to the red of 650 nm for copper and gold, the other two metals which show SERS at wavelengths in the 350–1000 nm region. The best morphology for surface plasmon resonance excitation is a small (<100 nm) particle or an atomically rough surface.

13.3. Measurements and Results

For the detection of explosives in air, a sensor surface with optimized, nanosize structures to optimize the Raman scattering will result in maximum amplification of the desired fingerprint signal. The best solution is to structure the Raman sensor in three dimensions, and take advantage of the depth of field information, or the information depth, of the technique.

Instead of receiving information from a surface layer, information from a three dimensional volume will be detected. This means that the sensitivity will increase drastically, i.e., instead of receiving information from a nanostructured surface layer, information from thousands of equivalent layers will be achieved, increasing the sensitivity considerably.

In a series of experiments we used membranes with nanometer sized metallic particles deposited on the membrane. As a reference sample a relatively high concentration TNT in methanol did not yield a visible spectrum when using an untreated membrane (no particles). Figure 2 shows experimental data from a membrane with 36 femtograms of TNT.

Similar measurements (Figure 3) were also made with RDX. The amount of RDX was 35 femtograms.

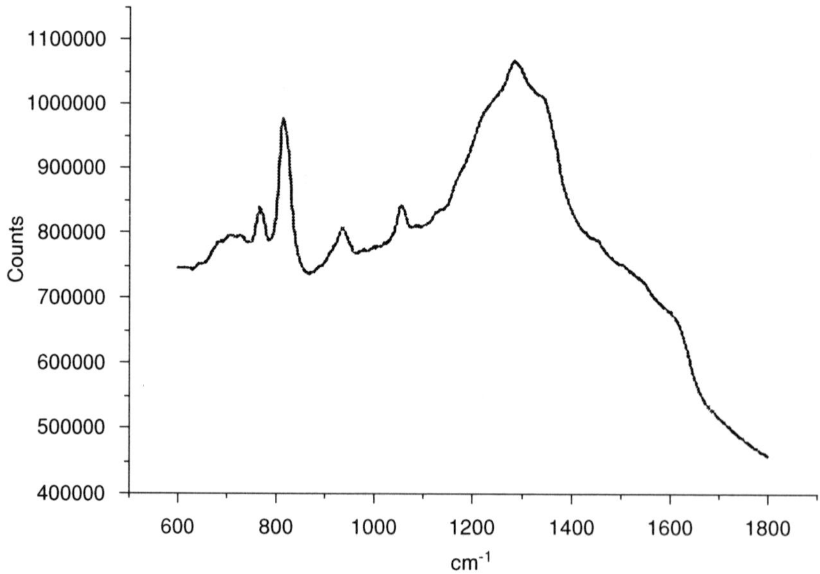

Figure 2. Raman spectrum of TNT using a Portendo sensor membrane. Analysed amount of TNT = 36 femtograms

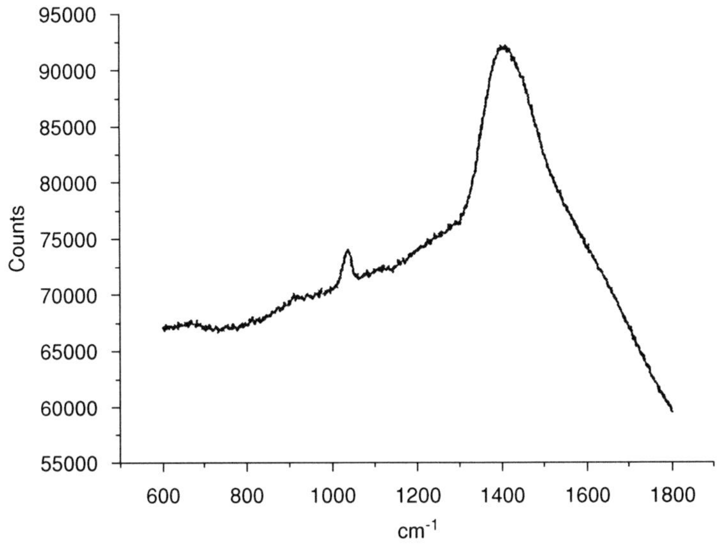

Figure 3. Raman spectrum of RDX using a Portendo sensor membrane. Analysed amount of RDX = 35 femtograms

13.4. System Technology

Portendo's TDS (Threat Detection System) is specified together with military expertise to address the highly qualified requirements set by anti-terrorist organisations. The requirements derive from solving the actual problem, rather than improving existing techniques. The result is a vision of a pervasive threat detection system that is easily deployable in a large number of different situations at a low cost. The TDS system compiles a large amount of information into a comprehensible picture of a threat situation. The major requirements are listed as follows:

(a) The system shall be able to simultaneously supervise large areas.
(b) The system shall aggregate information from several sensors into an overview picture.
(c) The system shall be operational 24/7 with real-time presentation.
(d) False alarm rate shall be low.
(e) System security shall be high.
(f) The system shall be harmless to both humans and the environment.
(g) Cost of deployment and ownership must be low.
(h) It should be possible to separate information recipients.

The TDS system implements these requirements using an operations central that collects information from sensors connected wirelessly or by wire. This configuration enables the TDS system to operate over large distances in real time. The sensor units, PSD's, operate autonomously, i.e. no human intervention is needed for their operation. Instead, application and intelligence expertise are working at the operations centre with a complete picture of the situation. This allows for low cost and rapid deployment of the TDS.

TDS communicates through commercially available radio networks; the PSD's are configurable with modules for different radio systems. For most land based applications, communications links provided by public mobile phone systems are sufficient. For sea based operations, PSD's can be equipped with additional communication modules such as VHF radio.

Due to the possible use of multiple communication subsystems, TDS implements security at a separate protocol layer above the transport layer in the protocol stack. This allows the system to operate the same security functionality regardless of communication method and removes any dependence of the communication subsystem to implement encryption, authentication and prevent eavesdropping.

With its modular security subsystem, the TDS implements information layering, a method of filtering different information types to separate end users. For example, a cargo transport company may use the TDS system for securing their containers used in overseas traffic. In reality, the transport company uses the security services provided by a security company since monitoring terrorist threats are of limited commercial interest. Instead, the cargo transport company use the logistic information provided by the TDS to monitor container locations and supervise cargo flows. By using the TDS information layering, commercially interesting information may be separated from security information for use at different destinations.

Finally, the TDS system allows users to start with relatively small sensor networks, allowing them to grow organically as demands change and new technologies arrive without large reinvestments. The TDS system can also integrate other functions such as logistics.

13.5. Conclusion

A distributed trace level explosives sensor network as described here represents a paradigm shift in approaching the threat from terrorism. The system allows early-warning capabilities and a wide-spread network of sensors, both which are lacking in other methods. This system combined with leading-edge explosives detection methods, as described earlier, allows this solution to severely diminish the terrorist threat.

14. TIME-RESOLVED EXCITATION–EMISSION MATRICES SPECTROSCOPIC APPROACH TO EXPLOSIVES DETECTION

S. Bobrovnikov
Institute of Atmospheric Optics, SB RAS, Akademicheskii Ave., 1, 634055 Tomsk, Russia

Abstract. The presentation is devoted to general analysis of the feasibility to use various effects of laser radiation interaction with the matter to detect traces of the explosives vapours in the atmosphere. Some problems in developing practical laser-based detectors are also discussed. Analysis presented gives information on the expected levels of vapours concentration in the atmosphere for the most common explosives and suggests possible scenario of their detection. Comparison is made in the presentation of the main optical effects regarding the efficiency of interaction between the matter and radiation and potentialities of the laser-based detection methods are being assessed. A possibility is analyzed of using for this purpose the fluorescence and resonance Raman scattering at excitation in deep UV region. To improve the selectivity of the detection it is proposed to make use of the spectroscopic approach based on the method of time-resolved excitation–emission matrices. This approach is capable of improving the detection selectivity by use of four-dimensional spectroscopic images of the objects surveyed. The way of realizing a detection system based on this approach is proposed that assumes development of a multichannel spectrum analyzer possessing high time resolution. As an example some preliminary results are presented on detecting minor molecular species in the atmosphere with a 32-spectral channel lidar. Some key research tasks are formulated in the presentation that are to be addressed in developing laser systems for detection of the explosives traces in the atmosphere.

Nowadays, great attention is being paid to development of the methods and instrumentation to detect traces of explosives in the atmosphere. This is a cruel necessity caused by numerous inhuman acts of terrorism, which are more and more often assassinated throughout the world. As a rule, the terrorist attacks are being made by suicide-bombers who are usually indistinguishable from other people in a crowd. In this connection, the methods of remote and detection of traces of explosives in air are of particular interest as those could provide a possibility of noncontact (secret) detection of the explosive or a bomber.

H. Schubert and A. Rimski-Korsakov (eds.), Stand-off Detection of Suicide Bombers and Mobile Subjects, 107–125.

Among many known methods of detection and identification of the chemical compounds, only the optical methods of making chemical analysis can provide the possibility of conducting remote measurements because they use, as the information carrier, light waves, either absorbed or emitted by the object to be detected. In the past decades, all optical methods, practically without exception, have undergone their second birth due to great progress in the laser and optoelectronics technologies. Thus, recent advances in semiconductor lasers, optical fibers, and high-sensitivity matrix detectors based on CMOS technology have made it possible to make Raman analyzers standard devices for routine measurements in chemical analysis of substances [1]. Lidar methods also gain a wide use in monitoring of the environment and in ecological control [2].

However, optical methods have so far found only limited application in detection of the explosives and many problems in the development of these methods are still to be addressed [3]. The difficulties in detecting the explosives are, first of all, caused by the extremely low concentration of the explosives' vapors in the atmosphere since these substances are low volatile. In this presentation we would like to attract attention to the following key aspects concerning the application of remote laser sensing methods to detection of the explosives:

1. What are the potentialities of the optical detection methods?
2. What are the conditions for their application to detection of the explosives?
3. Physical principles that makes the basis for optical detection methods comparison of the efficiency of the basic optical effects of interaction between light and matter.
4. Time-resolved exciting–emitting matrices approach for explosives detection.

14.1. Potentialities of the Optical Detection Methods

What are the potentialities of the optical methods in application to detection of chemical compounds in the atmosphere? In order to estimate the detectability of optical methods and to encourage instrumentation designers let us first assess the minimum detectable concentration under favorable conditions of no external noise and other interference factors. In order to estimate the detectability of optical methods and to encourage instrumentation designers we shall first estimate the minimum detectable concentration under favorable conditions of no external noise and other interference factors. To achieve this task consider how optical signals appear in the remote detection methods. Figure 1 depicts the optical arrangement, which illustrates the lidar detection and measurement principle.

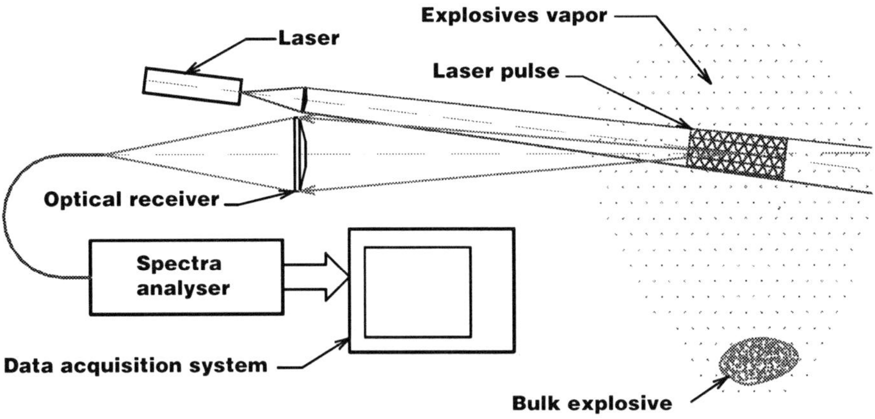

Figure 1. Block-diagram illustrating the operation principle of a lidar detector

The pulse of laser radiation interacts with the vapor of an explosive existing around the blasting assembly. The interaction yields an optical response, which is collected with an optical receiver. After spectral isolation and photo detection the detection signal is formed.

The main factors determining the signal power are as follows:

- mean power of the laser radiation;
- efficiency of the radiation interaction with the matter;
- solid angle within which the scattered radiation is being collected;
- efficiency of the receiving optics and electronics.

Taking into account the main factor listed above the mean number of photocounts per second recorded with such a generalized lidar detector can be estimated, accurate to an order of magnitude, by the following expression.

$$n_x = \frac{P_l}{h\nu} \cdot \sigma_x \cdot c_x \cdot \Delta r \cdot Q \cdot \frac{S_o}{r^2} \cdot \tau \cdot \eta, \tag{1}$$

where P_l is the mean power of laser radiation, $h\nu$ is the photon energy, σ_x is the absorption cross-section, c_x is the concentration of the vapor molecules, Δr is the spatial resolution, Q is the quantum yield of the reemission process, $\frac{S_o}{r^2}$ is the solid angle within which the scattered radiation is being collected, So is the area of the receiving aperture, r is the sounding range, τ is the transmission of the receiving optics, η is the quantum efficiency of the photodetector. The process of laser radiation transformation into a response signal from the medium is schematically shown in Diagram 1. According to this diagram the mechanism of transformation can approximately be quantified as follows.

$$n_l^{10^{18}} \xrightarrow[\text{absorption}]{\sigma_x \cdot c_x \cdot \Delta r = 10^{-12}} n_{abs}^{10^6} \xrightarrow[\text{emission}]{Q = 10^{-1}} n_{em}^{10^5} \xrightarrow[\text{propagation}]{\frac{S_0}{r^2} = 10^{-3}} n_{sig}^{10^2} \xrightarrow[\text{conversion}]{\tau \cdot \eta = 10^{-1}} n_x^{10}$$

Diagram 1. Scheme of the laser radiation transformation into a response signal

It is convenient to estimate the signal in the number of photocounts recorded. Thus a laser delivering radiation of 1-W mean power emits about 10^{18} photons per second (n_l in Diagram 1). Note in advance, that typical value of the absorption cross-section of the most of vapors in the near UV and visible range is $\sigma_X = 10^{-20}$ cm^2.

Let us take the concentration of the vapor to be at the level of $c_X = 10^7$ cm^{-3}, or 1 ppt in relative units under standard atmospheric conditions. We took the spatial resolution to be 10 cm. The values of the other parameters are shown in Diagram 1.

As follows from the diagram, 10^7 molecules of a substance per cubic centimeter will generate an optical response to the incident radiation of 1-W mean power sufficient to produce signal pulses at the output of the detection system at the count rate of 10 Hz, provided that the quantum efficiencies of the interaction and photodetection processes, as well as the transmission of the optical system are about 10%. One can show [4] that in the absence of noise it is quite sufficient to have 6 photocounts at the output of the detection system to reliably detect a signal with Poisson statistics. This means that in this, though slightly idealized case, we have a 1.5 times reserve in system sensitivity. To say generally, the above consideration shows that, theoretically, the inherently contactless and remote optical methods have quite high sensitivity and thus can successfully be applied to detection of the extremely low amounts of chemical substances.

One can find many experimental studies, which have become classical, in the literature devoted to detection and measurements of gases and vapors by use of optical methods [5]. Thus, Reference [6] reports on measurements of the OH radical concentration of 5×10^6 cm^{-3} as low. In Reference [7] a laser-induced fluorescence technique has been developed, which enabled measurements of the CN radical content at the level of 10^8 cm^{-3}. In the case of metal vapors characterized by much higher cross-sections of the interaction with light, the achieved threshold sensitivities of the detection using resonance fluorescence are even lower. Thus, for lead vapor the threshold concentration sensitivity was demonstrated at the level of 250 cm^{-3}, while for sodium vapor in the mesosphere it was at the level of several atoms per cubic centimeter [8, 9].

Thus, the sensitivity of optical methods estimated theoretically and demonstrated in practice is sufficiently high.

So, what are the specific conditions of applying the optical methods to detection of vapors of explosives in the atmosphere?

14.2. Conditions of Applying the Optical Methods to Detection of Vapors of Explosives in the Atmosphere

The task of detecting the presence of explosives in a volume under control by use of instrumental means is, as a rule, a part of security measures and therefore its achievement should meet the requirements to the security systems. Among these requirements there are: efficiency, reliability, remotability, secrecy, and harmlessness. However, in addition to these conditions it is very important to identify the object to be detected and to address the questions on what is the mean concentration of the explosives in vapor phase within the volume under control and what is the configuration of the vapor cloud and how the concentration of the vapor is distributed over the cloud. It is also useful to understand the range of concentration the instrument has to be sensitive to.

In Table 1 we summarize the requirements to the system intended for detecting the explosives.

Table 2 gives the list of most widely used explosives [10]. Of course this is not a complete list of substances, with which remote detection will deal in practice. From the standpoint of optical methods it is most important to have information on the molecular structure and on the type s of chemical bonds as those determine the spectroscopic properties of the chemical compounds.

Data collected in Table 3 give the information on the vapor density of some explosives in the atmosphere under conditions of nonrestricted vaporization [11]. These data can be used as a basis in developing he systems for detection of the explosives.

Thus we have formulated the conditions of application of the laser-based detectors of the explosives, the types of these chemical compounds and the level of vapor concentrations to be expected near blast assembly. It is quite pertinent to note that the concentration of explosives depends on the temperature and packing [12]. For this reason the above-listed estimates are quite tentative.

Consider now the physical principles, which make up the basis of the optical detection methods.

TABLE 1. Gerneral requirements to the system intended for detecting the explosives

Type of substance	Explosives
Phase state	Vapor
Concentration, ppt	0.1–10,000
Detection rate (s)	1–10
Range (m}	1–10
Special requirement	Secrecy

TABLE 2. The list of most widely used explosives

No.	Compound	No.	Compound
1	2,4,6-Trinitrotoluene (TNT)	11	Nitrozelluloze (NC)
2	Pentaerytritoltetranitrate (PETN)	12	1,3,3-Trinitroazetidine (TNAZ)
3	Hexahydro-1,3,5-triazine (RDX)	13	Nitroglycerin (NG)
4	Octahydro-1,3,5,7-tetranitro-1,3,5,7-tetrazine (HMX)	14	Picric acid (PA)
		15	Ammonium nitrate (AN)
5	Triacetone triperoxide (TATR)	16	Ammonium perchlorate (AP)
6	Hexamethylenetriperoxidediamine (HMTD)	17	Ammonium dinitramide (ADN)
7	Methyl-2,4,6-trinitrophenylnitramine (Tetryl)		
8	Ethylene glycol dinitrate (EGDN)	18	Potassium nitrate
9	Triaminetrinitrobenzene (TATB)	19	Potassium perchlorate
10	3-Nitro-1,2,4-triazol-5-one (NTO)	20	CI-20

TABLE 3. Data on the evaporability of some explosives

	Vapor density (order of magnitude)		
Compound	The number of molecules of the explosive per 10^{12} molecules of air (ppt)	The number of molecules of the explosive per 1 cm^3 of air	The mass of the explosive per 1 cm^3 of air (g)
NG	10^6	10^{13}	10^{-9}
TNT	10^4	10^{11}	10^{-11}
PETN	10^0–10^1	10^8–10^9	10^{-14}–10^{-13}
RDX	10^0	10^8	10^{-14}
C-4 (91% RDX + 9% of a sticker)	10^{-1}	10^7	10^{-15}
EGDN	10^8	10^{15}	10^{-7}

14.3. Physical Principles, which Make Up the Basis of the Optical Detection Methods

All the optical methods for determining chemical composition of the ambient air employ the basic phenomena of interaction between light and the matter. It is only important that energy exchange occurs between the incident electromagnetic field and molecules of a substance during an interaction. In this case the reemitted wave will bear information on the structure of energy levels in a molecule. It is just the uniqueness of the energy levels structure of each molecular species that forms the fundamental principle of the recognition of chemical compounds.

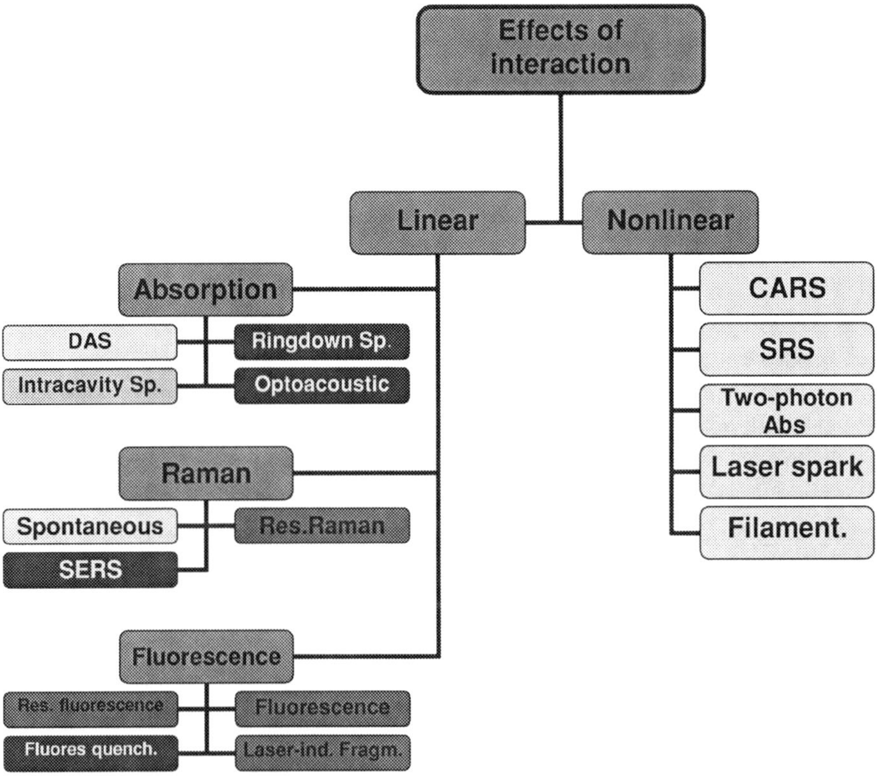

Figure 2. Diagram of the effects of interaction between the matter and radiation

To better understand the phenomenological basis of the optical detection methods it seems to be useful considering the whole set of effects of interaction between the matter and radiation from the stand point of their use in practice. Figure 2 shows a diagram of such effects, which allows one to analyze the practicality of their use in achieving the task of detecting the explosives.

In composing this diagram, we divided the effects into two groups, the linear and nonlinear ones. As to the linear effects those have been known long ago, while the study of nonlinear effects have only become possible due to advent of lasers, the light sources capable of emitting coherent radiation of high-power density.

The linear interaction effects have found wide-ranging application in chemical analysis as a basis for various analytical technologies.

From the linear effects the diagram in Figure 2 presents three main interaction effects, namely light absorption, spontaneous Raman scattering of light, and fluorescence. Those are just the effects that made the basis for the majority of analytical techniques developed so far.

The absorption is the effect of interaction between radiation and the mater that results in that a portion of radiation energy is absorbed by an atom or a molecule, which then can be transformed into other forms of energy, for example, into heat. Each molecular species has a unique absorption spectrum. The absorption spectrum can be obtained in the form of a transmission spectrum, the reflectance spectrum, the spectrum of attenuation of a return signal (in lidar technique), as well as in the form of spectrum of optical losses of a resonator, and so on. Analysis of the absorption spectra enables one to identify the species absorbing the light. Absorption spectra of various substances recorded in the infrared wavelength region are normally most informative since these are formed due to unique combinations of the molecular vibrational a rotational motions of the majority of substances. Although the technology of IR-spectroscopy is being permanently improved there still are many problems to be addressed in order to make use in full measure the information potential of the IR region. Table 4 presents information on the typical threshold sensitivities of different methods, which employ the effect of light absorption. As seen from Table 4 all these methods provide for a high sensitivity, but only one of them, namely the DAS technique, is a remote one.

However this technique gives good results only for the case of long-path absorption, for which reason it has found practical application only in ecological monitoring of the environment.

The effect of spontaneous Raman scattering of light can briefly be characterized as the interaction process of inelastic scattering of photons by molecules of a substance. This process results in a change of the energy state of interacting molecules and creation of photons having energy different than that of incident ones [13]. Raman spectra consist of bands shifted with respect to line of exciting radiation on the frequency scale. The spectral shifts of the bands are determined by the energy of vibrational-rotational motions of a molecule. As in the case of absorption spectra, Raman spectra also have high information content. However, in contrast to the absorption spectra, the positions of Raman spectra on the wavelength scale are determined by the wavelength of exciting radiation, which can be optimized when necessary.

TABLE 4. Detectability thresholds of the absorption methods

Method	Detectability threshold	Sampling needed, yes/no
DAS	ppb	No (long paths)
Intracavity spectroscopy	ppb	Yes
Ringdown Sp.	ppb	Yes
Photoacoustic spectroscopy	ppb	Yes

TABLE 5. Detectability limits for Raman scattering methods

Method	Threshold sensitivity	Sampling need, yes/no
Spontaneous Raman	ppm	No
Resonance Raman	ppb	No
Surface enhanced Raman scattering (SERS)	ppt	Yes

In Table 5 we present detectability thresholds for Raman methods [14]. It is seen from this Table that the sensitivity of Raman methods is in sufficient for detecting traces of the explosives in the atmosphere.

At the same time, if the frequency of exciting radiation approaches an allowed transition of a molecule the efficiency of Raman scattering undergoes resonant increase that can achieve about four orders of magnitude and even higher [15]. Therefore, resonance Raman technology can be a good candidate for use in detection of the explosives, being inherently contactless and remote technique.

The fluorescence is a classical phenomenon of optical interaction that has found wide-ranging application in practice. The fluorescence effect consists in that the molecule or an atom excited due to absorption of incident radiation can emit photons while relaxing from an excited to the ground state [16]. Actually, it does not mater how the molecules or atoms are excited, but in the case of excitation by laser radiation the possibility appears of conducting remote spectral analysis of media. Typically, fluorescence has high quantum yield. This is explained by the interaction between incident radiation and the matter occurs resonantly. It is for this reason that the fluorescence methods have high sensitivity, while being remote techniques at the same time.

Table 6 gives the data on the threshold detectability of the fluorescence methods [5]. It is obvious that such sensitivity well fits the requirements to the methods of detecting the explosives. For this reason the fluorescence deserves special attention as a candidate in developing highly sensitive detectors of the explosives and their traces in the atmosphere.

TABLE 6. Threshold values of the detectability of fluorescence methods

Method	Detectability limit	Sampling need, yes/no
Molecular fluorescence	ppt	No
Resonance fluorescence	10^2 cm^{-3}	No
Fluorescence quenching	ppt	Yes

Thus we have briefly overviewed the possibilities of using linear effects of interaction between the matter and light in constructing methods for remote detection of explosives in the atmosphere. It seems so that, in principle, fluorescence and resonance Raman scattering possess the highest sensitivity among the optical methods that can be developed for remote detection of explosives.

A detailed analysis of nonlinear effects of optical interactions is out of the scope of this presentation. However, it is worthy to note that significant advances in the equipment available, particularly femtosecond lasers, promises some new options for using nonlinear effects in remote explosives detection.

14.4. Time-Resolved Excitation–Emission Matrices Approach for Explosives Detection

Thus, analysis of the approaches to remote detection of the explosives' vapors in the atmosphere has shown that resonance Raman scattering and fluorescence of the explosives themselves and/or their fragments are the most promising ones. However, for all that the methods based on the fluorescence effects provide higher sensitivity of the detection. At the same time one should be careful as the spectra of fluorescence of polyatomic molecules comprise, as a rule, much complicated bands of spectral lines, the structure of which reflects huge number of degrees of freedom in such molecules. Assignment of such spectra makes a host of problems and calls for application of highly specialized techniques in order to achieve selectivity in detecting the species targeted.

Among the assignment, or recognition, methods the method of excitation–emission matrix diagrams (EEM) is quite an effective means. This fluorescence-based approach has found wide range of application in ecology and medicine. The basic principle of the method is in combining the excitation and emission spectra of an object under study in a single matrix. In this case the spectral response of the object is a function of two variables set on the plane (λ_{EX}, λ_{EM}), where λ_{EX} and λ_{EM} are the wavelengths of the exciting radiation and the emission, respectively. Thus, this response makes up a three-dimensional spectroscopic image of a species. An example of the EEM diagram of a biological object in the atmosphere is presented in Figure 3. One can also see, in this diagram the responses due to Rayleigh and Raman scattering of exciting radiation by the atmospheric components. The EEM diagram presented in Figure 3 just illustrates the possibilities of distinguishing among the spectral images of different species using such matrices.

The approach based on matrix representation enables one to present most comprehensively the spectroscopic properties of the objects studied as it shows

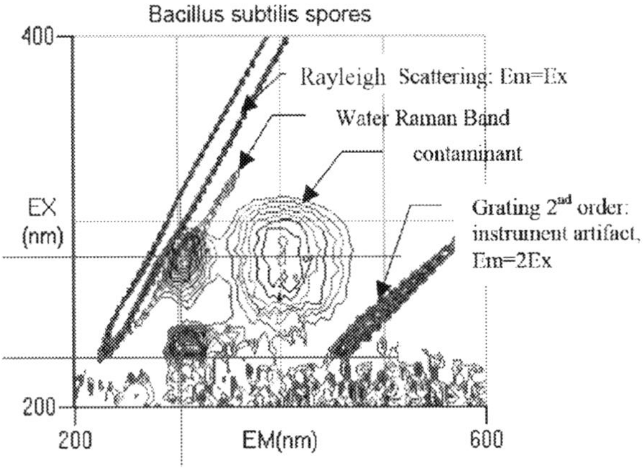

Figure 3. Example of excitation–emission matrix diagram [17]

evolution of the relevant spectral responses depending on the wavelength of exciting radiation.

If exciting a spatially localized object with pulsed radiation its spectral image undergoes time variation and this can favor increasing the detection selectivity.

Time-resolved spectral response produced at excitation by radiation with the wavelength λ^i_{EX} can be described by a function of two variables $\Phi_i(\lambda_{EM}, t) = F(\lambda^i_{EX}, \lambda_{EM}, t)$ that represents it as a 3-D image bearing information on the specific properties of the object to be detected. If, at the same time, the time-resolved responses are obtained using exciting radiation at a set of wavelengths λ_{EM}, then the object under study will already be presented a four-dimensional characteristic image $R = F(\lambda_{EX}, \lambda_{EM}, t)$. It is obvious that a four-dimensional image bears more information on the characteristic properties of the objects detected in this way.

From the standpoint of methodology, four-dimensional spectral images more comprehensively describe the spectroscopic properties of a system and, therefore, the technique based on recognition of such images would provide achieving the best possible detection selectivity.

The characteristic four-dimensional image $R = F(\lambda_{EX}, \lambda_{EM}, t)$ is nothing but a time-resolved EEM diagram. Figure 4 shows a hypothetical view of such a matrix simulated numerically for several moments in time.

Obtaining four-dimensional spectroscopic responses of the objects to be detected assumes the excitation of the object to be made successively or simultaneously using radiation of different wavelengths from some informative range of wavelengths. In so doing one has to record a time-resolved spectral response to exciting radiation at each of the wavelengths used. The set of all these spectral responses will make up the sought four-dimensional image of

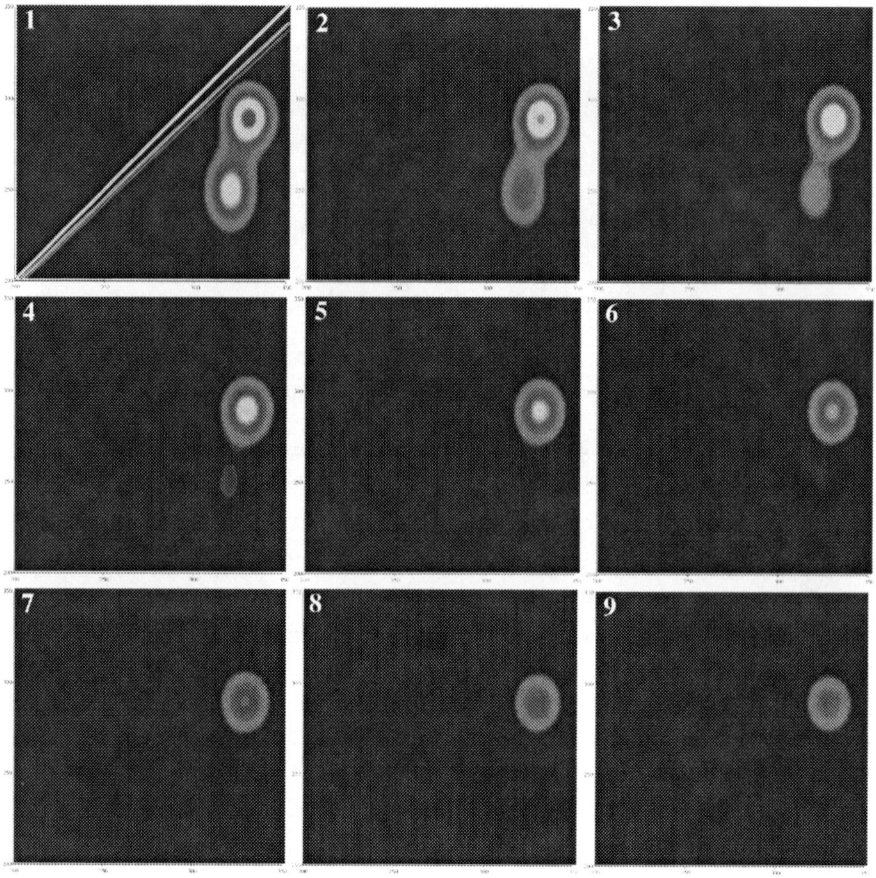

Figure 4. Simulated view of the time-resolved EEM diagram

the object sounded. Thus, it is clear that a four-dimensional spectral image may have high information content on the object under study. However, acquisition of such information requires much energy and time. It is for these reasons that success in realization of this approach in practice can only be achieved by use of up-to-date multichannel spectroscopic technologies in combination with high-power tunable lasers.

In recent years, intense development has been carried out at the Institute of Atmospheric Optics of high-speed multichannel recorders of light signals based on the use of multianode photomultipliers. Thus, a 32-channel photon counting system developed for use with a 32-anode H7260 Hamamatsu PMT has been used in several lidar systems for studying the atmospheric water vapor and liquid water in clouds by use of the rovibrational $\nu 1$ Raman band of this molecule [18]. Use of the multianode PMTs enables one to achieve

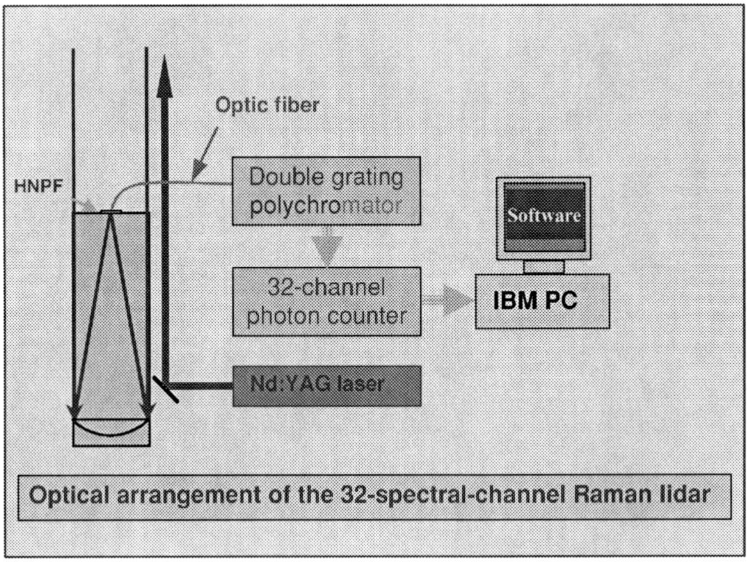

Figure 5. Block-diagram of a 32-channel Raman lidar

time resolution of measurement about 1ns and sensitivity at the mean level of thousandths of a photon per second. High-speed spectrum analyzer, if built up around such a multichannel photon counting system enable one to simultaneously acquire 32 lidar returns with the spatial resolution of 1.5 m each. Block-diagram of a 32-channel Raman lidar for remote measurements of liquid water content in clouds is depicted in Figure 5.

Appearance of the 32-channel Raman lidar built up in cooperation with the lidar group at Max Planck Institute for Meteorology, Hamburg, Germany is shown in Figure 6. A 32-channel lidar response acquired with this lidar is a 3D spectral image, which can easily be presented using means of 3D graphics on a computer. Figure 7 presents an example of a 3D spectral image a laser pulse (pulse of laser sounding radiation). Horizontal axes show channel numbers and time, while the vertical axis shows the response intensity.

Figure 8 presents a time-resolved spectral response from an orange-colored glass irradiated by laser radiation at 532 nm wavelength. One can easily see from this figure a 3D image of the exciting laser pulse (green peak in the center) and a 3D image of a fluorescence emission from the color glass. It is obvious that 3D presentation enables one to improve the reliability of distinguishing lidar returns.

Figure 9 shows a spectral image of the rovibrational Raman band of water vapor and liquid water acquired at vertical sounding of a cloud base with a 32-spectral-channel lidar. Like in the previous case, analysis of the 3D image

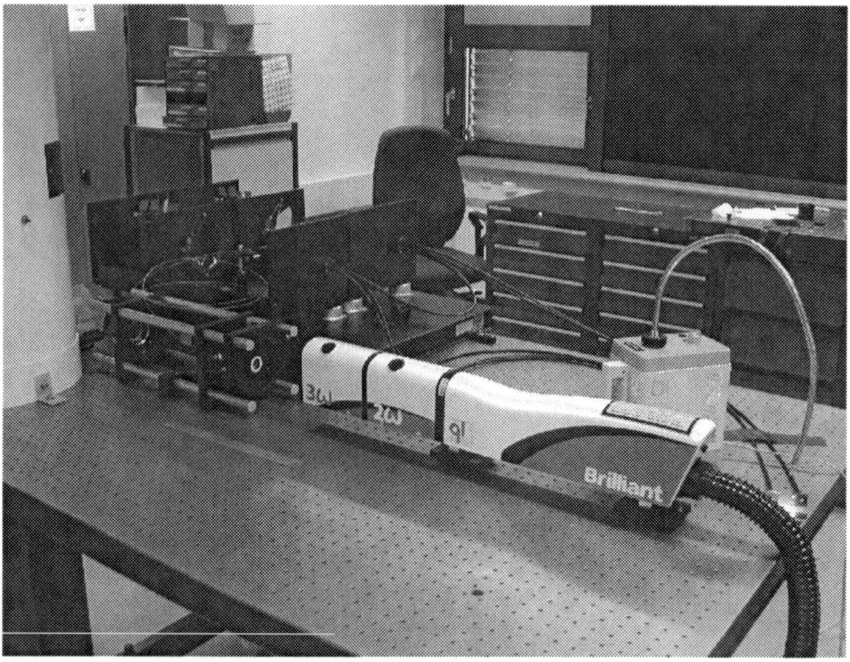

Figure 6. Appearance of the 32-channel Raman lidar built up in cooperation with the lidar group at Max Planck Institute for Meteorology, Hamburg, Germany

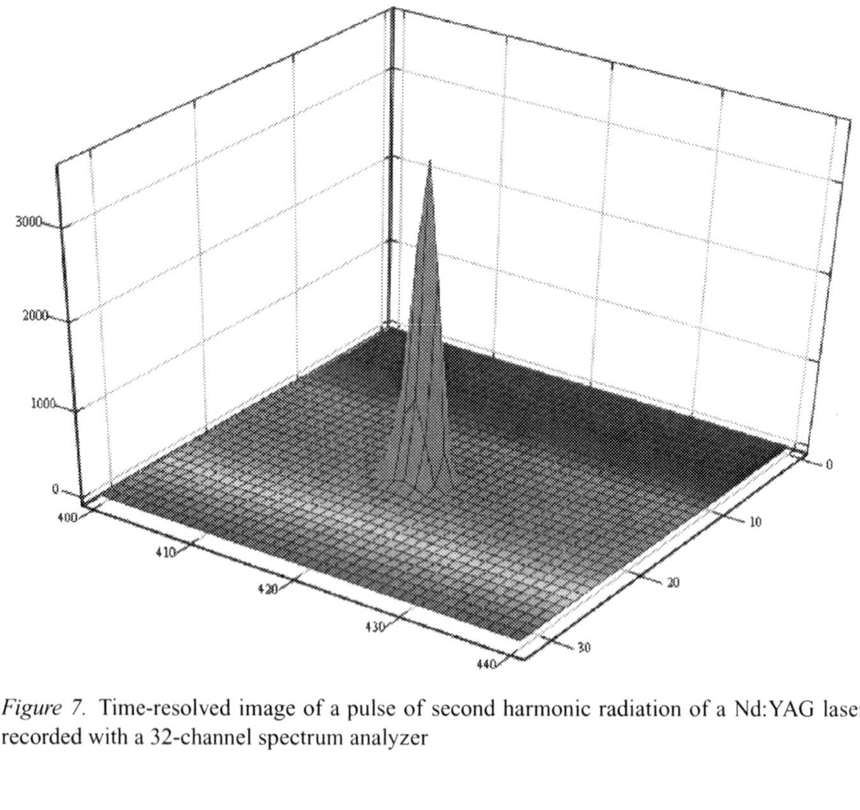

Figure 7. Time-resolved image of a pulse of second harmonic radiation of a Nd:YAG laser recorded with a 32-channel spectrum analyzer

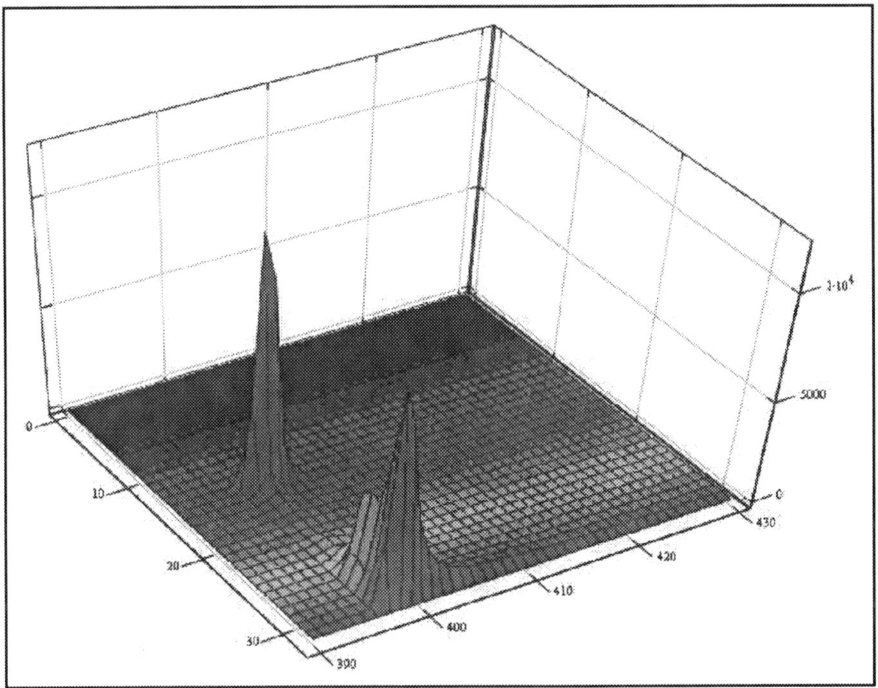

Figure 8. Time-resolved spectral response from an orange-colored glass irradiated by laser radiation at 532 nm wavelength. One can easily see from this figure a 3D image of the exciting laser pulse (green peak in the center) and a 3D image of a fluorescence emission from the color glass

enables one to detect the presence of liquid water in the atmosphere by its spectral and time image in the cloud base.

Until now, we considered a spectral-temporal 3D image of an object sounded using sounding radiation at a single wavelength. As is seen from quite general considerations and having in mind the examples given, even single frequency excitation of the fluorescence the multichannel time-resolved spectroscopic approach proposed is quite an effective means of distinguishing among the objects sounded. If this approach is incorporated into the technology of multifrequency excitation that, for example, is being used in generating EEM diagrams then one obviously obtains a new powerful instrument for detecting and identification of various fluorescing objects, which uses comprehensive information on spectroscopic properties of the objects sounded. Block-diagram of such an instrument is depicted in Figure 10.

The beam of a frequency tunable pulsed radiation is directed toward an object to be sounded. The optical response occurring due to interaction between laser radiation and matter is collected with an optical receiving system

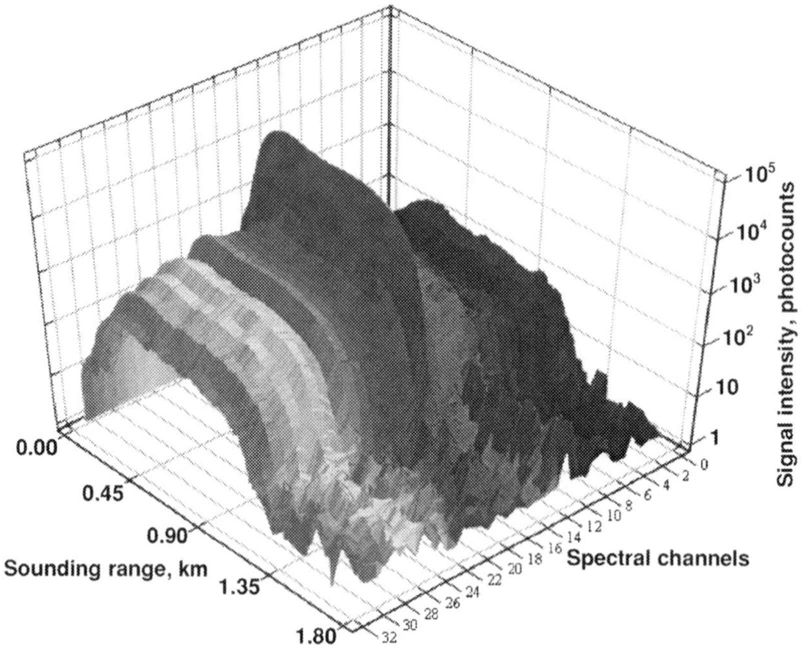

Figure 9. Spectral image of the rovibrational Raman band of water vapor and liquid water acquired at vertical sounding of a cloud base with a 32-spectral-channel lidar

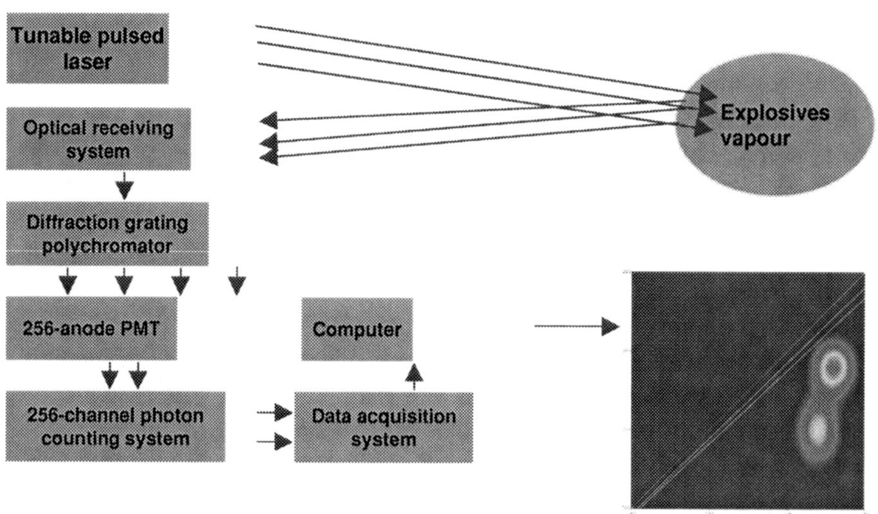

Figure 10. Block-diagram of a laser detector of explosives in the atmosphere based on four-dimensional spectral imaging of the objects sounded

Figure 11. A 256-anode H9500 PMT assembly; Hamamatsu Photonics

and then radiation enters into a multichannel spectrum analyzer. Time behavior of the response intensity is recorded in each of the spectral channels isolated. Such measurements are being repeatedly carried out for each of the wavelengths of the exciting radiation chosen within the tuning range. Measurement data are entered into a computer for the subsequent analysis and interpretation. Each of the measurement cycles yields a spectroscopic image of the object.

We suggest to increase the number of spectral channels of the analyzer from 32 to 256 by use of a newly developed at Hamamatsu Photonics 256-anode matrix PMT assembly H9500 (see Figure 11). Thus enhanced spectral resolution will provide necessary level of the system selectivity in analyzing multispecies objects.

14.5. Conclusion

In conclusion, we should like to note that despite extremely tough conditions for application optical methods are quite promising for achieving the task of operatively detecting explosives' traces in the atmosphere. Among those the methods based on resonance emission effects like Raman scattering and fluorescence are most sensitive. Development of these detection techniques is yet at early stages and thus needs many additional investigations to be carried out of the spectroscopic properties of the explosives' vapors as well as of the ways to efficiently excite the explosives' molecules and their fragments by optical radiation. Along with the laser fragmentation methods, which are most promising now, it is also worth developing the detection methods that are based in the effects of direct interaction between laser radiation and complex chemical compounds. The success of such studies strongly depends on the advances in modern spectroscopic methods. In this connection, the method of time-resolved EEM diagrams attracts much attention as an absolute spectroscopic approach enabling one to acquire most complete information on the capability of an object under study to respond to optical action within some interval of wavelengths (the four-dimensional image). Successful realization of this method is mostly determined by the advantages in two branches of technology, namely tunable lasers and fast operating multichannel spectrum analyzers. Both of these technologies have been rapidly developed in recent times. We have also contributed, so modestly, to this research field and hope that this will help development of remote systems for detecting explosives in the atmosphere.

References

[1] Christesen, S., M. Carrabba, and J. Bello, 2002. Nonintrusive Raman analysis of chemical agents. Vibrational Spectroscopy-Based Sensor Systems, Proceedings of SPIE—The International Society for Optical Engineering, 4577, pp. 287–293.

[2] Balin1, I., V. Simeonov, I. Serikov, S. Bobrovnikov, B. Calpini, Yu. Arshinov, and H. van den Bergh, 2004. Simultaneous measurement of temperature, water vapor, aerosol extinction and backscatter by Raman lidar, Reviewed and Revised Papers Presented at the 22nd International Laser Radar Conference (ILRC 2004), 12–16 July, 2004, Matera, Italy, edited by Gelsomina Pappalardo, Aldo Amodeo.

[3] Arusi-Parpar, T., D. Heflinger, and R. Lavi, 2001. Photodissociation followed by laser-induced fluorescence at atmospheric pressure and 24 °C: A unique scheme for remote detection of explosives, Appl. Opt., 40, 6677–6681.

[4] Arshinov, Yu., S. Bobrovnikov, I. Serikov, A. Vorozhtsov, and N. Eisenreich, 2005. Remote detection of chemical agents with a lidar in the solar-blind spectral region, 36th International Annual Conference of ICT & 32nd International Pyrotechnics Seminar 28 June–1 July, 2005 Karlsruhe Federal Republic of Germany.

[5] Bolshakov, A.A., and S.V. Oshemkov, 1983. Flourescence gas analysis, in Analtsis of Inorganic Gases, Leningrad, Nuaka, pp. 28–36.

[6] Wang, C.C., and L.I. Davis, 1974. Rev. Lett., 32, 349.

[7] Jackson, W. M., 1973. J. Chem. Phys., 59, 960.

[8] Gelbwachs, J.A., C.F. Klein, and J.E. Wessel, 1977. Appl. Phys. Lett., 30 (N9), 489.

[9] Bolshov, M.A., A.V. Zybin, and V.G. Koloshnikov, 1980. Quantum Electron., 7 (No 8), 1808.

[10] Application Guide AG-005 Raman Detection of Explosives, Tienta Sc., Inc., www.tientasciences.com.

[11] Vandyshev, B.A., 1998. "Explosives detection by analyzing their vapors and particles" Special-Purpose Technology, No. 1.

[12] Mostak, P., 2004. Vapour and trase detection of explosives, Vapour and Trase Detection of Explosives for Anti-Terrorism Purposes, Kluwer Academic Publishers, Dordrecht, Boston, London.

[13] Introduction to the Theory of the Raman Effect by J.A. Koningstein, 1972. D. Reidel Publishing Company, Dordrecht, Holand.

[14] Long, D., 2001. The Raman Effect, John Wiley & Sons, Ltd.

[15] Weber, A. (Ed.), 1979. Raman Spectroscopy of Gases and Liquids, Springer, Berlin, Heidelberg, New York.

[16] Fluorescense and Phosphorescence by Peter Pringsheim, 1949. New York, London.
[17] www.photonsystems.com.
[18] Arshinov, Yu.F., S.M. Bobrovnikov, A.I. Nadeev, I.B. Serikov, D. Kim, H. Cha, and K. Song, 2002. Observation of range-resolved rovibrational Raman spectra of water in clear air and in a cloud with a 32-spectral-channel Raman lidar Twenty-First International Laser Radar Conference (ILRC21), Québec, Canada, 8–12 July, 2002.

15. MID-INFRARED LIDAR FOR REMOTE DETECTION OF EXPLOSIVES

Christoph Bauer[1], Jörg Burgmeier[1], Christian Bohling[2], Wolfgang Schade[1,2], and Gerhard Holl[3]
[1]*LaserAnwendungsCentrum, Technische Universität Clausthal, Arnold Sommerfeld Str. 6, 38678 Clausthal-Zellerfeld, Germany*
[2]*Institut für Physik und Physikalische Technologien, Technische Universität Clausthal, Leibnizstrasse 4, 38678 Clausthal-Zellerfeld, Germany*
[3]*Wehrwissenschaftliches Institut für Werk-, Explosiv- und Betriebsstoffe (WIWEB) Grosses Cent, 53913 Swisttal, Germany*

Abstract. Laser ablation spectroscopy combining with high-resolution mid-infrared absorption measurements offers interesting possibilities for sensitive and selective stand-off analysis of surface contaminations under real-time operation conditions. The detection of NO_x production rates emitted from surfaces after interaction with a pulsed infrared laser beam allows distinguishing between different surface contaminations, e.g. energetic and non-energetic materials but also between molecules with similar atomic composition. This is shown for the explosives TNT, Octol and HMX. For efficient laser fragmentation of surface contaminations the excitation wavelengths 1.06 μm and 1.47 μm are compared indicating that a pulse power as low as 0.25 mJ/pulse at 1.47 μm is sufficient for laser-induced surface evaporation of explosive contaminations. Then no plasma is ignited by the infrared laser pulses and therefore the obtained NO_x emission is only due to the explosive surface contamination.

15.1. Introduction

The atomic composition of various organic materials is given by the elements carbon (C), nitrogen (N), oxygen (O) and hydrogen (H). When focusing an intense pulsed laser beam on the surface to be investigated, a plasma will be generated instantaneously by the light matter interaction.

[1]http://www.lac.tu-clausthal.de
[2]http://www.pe.tu-clausthal.de/AGSchade

H. Schubert and A. Rimski-Korsakov (eds.), Stand-off Detection of Suicide Bombers and Mobile Subjects, 127–133.

Surface molecules are fragmented into their atomic compounds and excited atoms, molecules and molecular compounds as well as electrons are generated in the plasma. When the laser pulse is off no energy is transferred to the sample anymore and the ions and electrons recombine as well as the excited states which results in the emission of light. Spectrally-resolved detection of the plasma emission gives information on the atomic composition of the sample ("first order fingerprint"). This can be used for characterization of different samples if the atomic composition is different from sample to sample. As an example, plastics and explosives are composed by the same elements and therefore cannot be classified by only measuring laser-induced plasma intensities. However, when additionally using the temporal evolution of the plasma emission for specific wavelengths a data analysis can be developed that allows a more specific sample analysis. An example is shown in Figure 1. Here laser-induced plasma spectroscopy (LIBS) is applied to distinguish between an explosive and an organic (wood) sample. Analyzing the LIBS intensity spectra emissions from N, O, H, Na and K but also

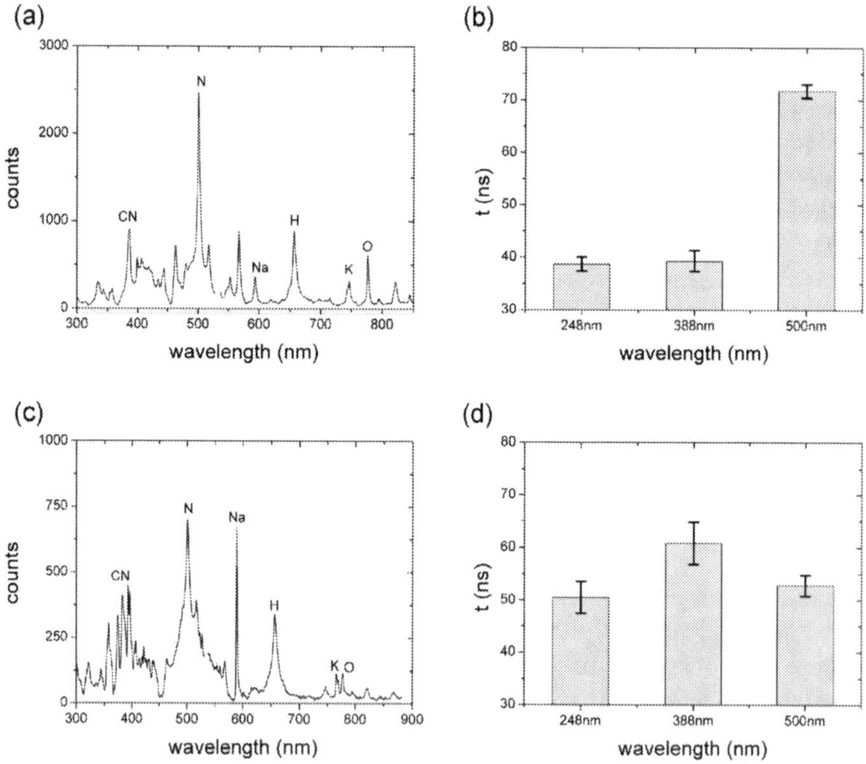

Figure 1. LIBS intensity spectra (a, c) and time resolved data analysis of LIBS spectra for specific emission lines (b, d) for the samples TNT (a, b) and wood (c, d)

from molecular fragments such as cyanide (CN) are obtained (Figure 1(a) and (c)) and the LIBS intensity spectra of both samples are very similar. The strong carbon emission is because of the relatively high amount of carbon in these materials but the CN emission is explained by a reaction of the carbon emitted from the material and the nitrogen also from the sample but additionally from the ambient air which is also ionized by the plasma. However, closer inspection of the time-evolution of the C-, CN- and N-emission at 248, 388 and 500 nm, respectively, allows clear classification of the two samples.

Depending on the absorption cross section for the materials under investigation laser-induced plasma temperatures are different for different materials. Plasma temperatures of up to 12,000 K are obtained when using laser pulses of several millijoules.

This effect results in characteristic LIBS intensity decay times for different materials if the laser excitation is performed under same experimental conditions. The corresponding decay times are typically in the order of several tens to a few hundreds of nanoseconds ("second order fingerprint") depending on the applied laser power, the excitation wavelength and the material to be analyzed. Since LIBS is known as a contact-free diagnostic method it found several practical applications in industrial process control, environmental analysis and very recently also for security applications [1–6]. However, to improve the LIBS technique as an analytical stand-off tool with respect to more selectivity especially for analyzing surface contaminations the combination of LIBS with other laser diagnostics such as mid-infrared (MIR) absorption spectroscopy has to be investigated. In this context selective and sensitive NO_x detection after the laser ablation or evaporation process indicates a very high potential for analyzing explosive residuals on surfaces [7]. To achieve only a surface and not a bulk material interaction with the ablating laser pulse it is important to adjust the laser wavelength to the absorption coefficient of the material to be analyzed. In case of detecting contaminations of explosive on metals it turned out that use of 1.5 μm laser wavelength instead of the conventional Nd:YAG laser wavelength 1.06 μm is superior for the laser ablation/fragmentation (PLF) process. Only 0.25 mJ pulse energy at 1.5 μJ are sufficient for plasma generation or efficient PLF on explosives or plastics while more than 1 mJ are necessary using the 1.06 μm. Then only laser interaction takes place with the surface contamination (explosive) but not with the bulk material (metal). These results offer new and interesting possibilities for developing stand-off diagnostic tools applying miniaturized passively Q-switched microchip Er:Glass lasers as seed laser for fiber amplifiers that are frequently used in telecommunications for LIBS or PLF. The NO emission of the material to be investigated is simultaneously measured by high resolution absorption spectroscopy applying difference-frequency generation

(DFG) or quantum cascade (QC) lasers operating around 5 μm [8, 9]. First results are discussed in the present paper.

15.2. Experimental Setup

Remote stand-off detection of explosives is proposed by combination of pulsed laser fragmentation (PLF) and pulsed mid-infrared differential absorption spectroscopy (MIR-DIAS). The proposed experimental set-up is shown in Figure 2.

A pulsed laser excitation with a wavelength above 1.5 μm induces a heating and subsequently fragmentation (PLF) of surface contaminations to be investigated. A second with respect to the PLF laser delayed MIR laser pulse measures the NO concentration that is generated by laser-material interaction on the surface to be investigated. The sample itself is used as retro-reflector or scattering shield and since fragmentation or heating is only obtained by the 1.5 μm laser pulse on the surface of the sample to be analyzed a remote detection scheme is achieved by this method when measuring the reflected or backscattered MIR light by a telescope and MIR detector. A 1.5 μm laser excitation compared to the conventional 1.06 μm Nd:YAG laser line is preferred because of much stronger absorption of infrared shifted wavelengths due to the enhanced infrared absorption coefficient of explosives. Then the laser-material interaction is concentrated on the explosive surface contamination and not by the laser bulk material interaction which also induces NO_x under ambient air conditions and which are much stronger than the very weak NO_x signals generated by the explosive-laser light interaction if concentrations of

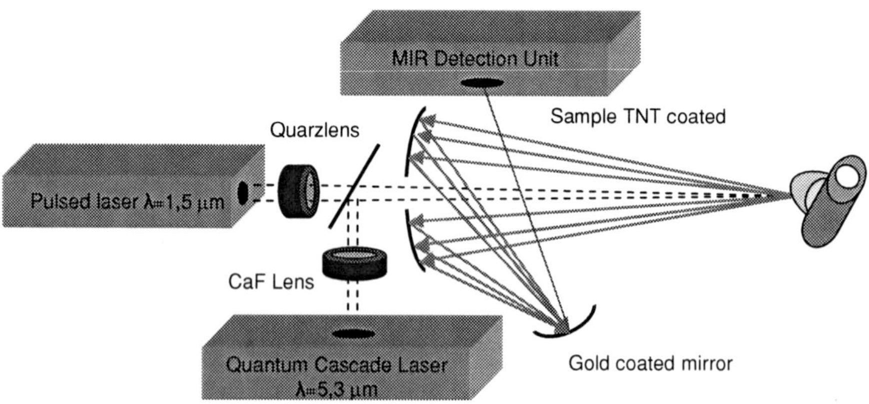

Figure 2. Stand-off detection of explosive contaminated surfaces by infrared PLF followed by MIR-DIAL spectroscopy.

several μg have to be analyzed. Applying a fiber amplified microchiplaser for PLF and a quantum cascade laser (QCL) operating at 5.28 μm for NO detection a stand-off operation distance of 10 m is possible.

15.3. Results and Discussion

Laboratory measurements indicate the possibility to distinguish between energetic and non-energetic materials by measuring the amount of emitted NO_x after PLF on the surface of a sample to be investigated, especially when calculating the ratio NO/NO_2. Because of the very low laser power at 1.5 μm there is no plasma ignition on a metal sample and therefore no NO_x signal can be detected. However, because of a much higher absorption of plastic materials at 1.5 μm compared to 1.06 μm, even pulse powers as low a 0.25 mJ/pulse at 1.5 μm induce a PLF that can be detected by measuring NO_x emitted from such surface. The same is true for an explosive contaminated metal surface. There is no significant PLF signal from the metal but because of the enhanced absorption at 1.5 μm for TNT there is a significant interaction with the TNT on top of the metal that results in NO_x emission that can be monitored. It is interesting to note that the ratio NO/NO_2 is completely different for nonenergetic (plastic) and energetic (TNT) materials. Therefore, this ratio allows to distinguish between explosive and non-explosive surface contamination (see Figure 3).

On the other hand, the nitrogen monoxide (NO) molecule shows fundamental vibrational absorption lines in the spectral range above 5 μm with oscillator strengths that allow sensitive spectroscopic diagnostics of NO. A pulsed laser excitation with a wavelength above 1.5 μm induces a heating and subsequently vaporization of surface contaminations that have to be investigated. A second with respect to the PLF laser delayed MIR laser pulse measures the NO concentration that is generated by laser-material interaction in a small volume. The sample itself is used as retro-reflector or scattering

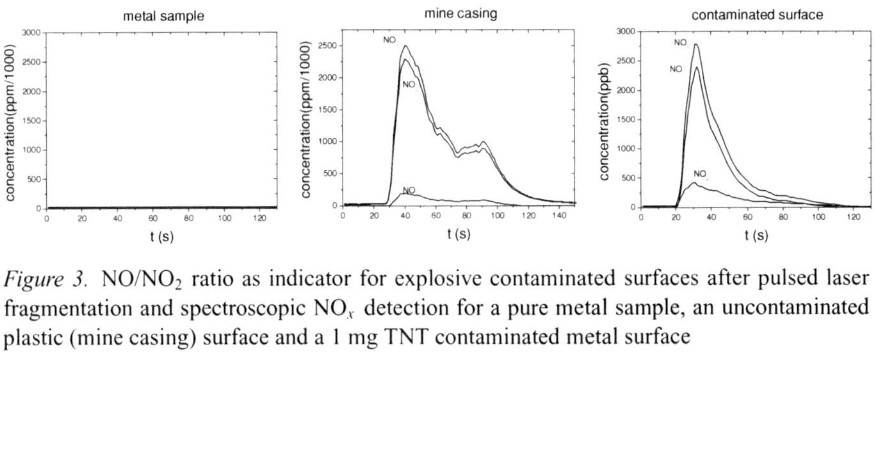

Figure 3. NO/NO_2 ratio as indicator for explosive contaminated surfaces after pulsed laser fragmentation and spectroscopic NO_x detection for a pure metal sample, an uncontaminated plastic (mine casing) surface and a 1 mg TNT contaminated metal surface

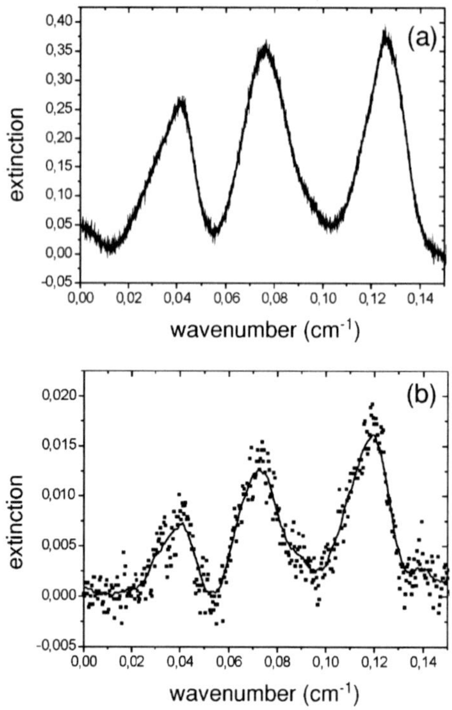

Figure 4. NO absorption spectrum detected by a QCL at 5.28 μm after surface fragmentation of a HMX sample by interaction with infrared laser radiation at 1.5 μm. (a) NO reference cell. (b) PLF and MIR-DIAS under ambient air conditions

shield and since fragmentation or heating is only obtained by the 1.5 μm laser pulse on the surface of the sample to be analyzed a remote detection scheme is achieved by this method when measuring the reflected or back-scattered MIR light by a telescope and MIR detector. In this concept 1.5 μm laser excitation compared to the conventional 1.06 μm Nd:YAG laser line is preferred because of much stronger absorption of infrared shifted wavelengths due to the enhanced infrared absorption coefficient of explosives. Then the laser-material interaction is concentrated on the explosive surface contamination and not by the laser bulk material interaction which also induces NO_x under ambient air conditions and which are much stronger than the very weak NO_x signals generated by the explosive-laser light interaction if concentrations of several μg have to be analyzed.

In first experiments a fiber amplified microchiplaser for PLF and a quantum cascade laser (QCL) operating at 5.28 for NO detection are applied. The NO emission was detected via QCL absorption spectroscopy when a HMX sample was heated by a sequence of infrared laser pulses at 1.5 μm. A typical NO spectrum from such measurements is shown in Figure 4.

The combination of these two techniques provides a very sensitive, miniaturized and efficient eye-safe stand-off technique for the analysis of trace explosive contaminated surfaces.

References

[1] Sattmann, R., I. Mönch, H. Krause, R. Noll, S. Couris, A. Hatziapostolou, A. Mavromanolakis, C. Fotakis, E. Larrauri, and R. Miguel, 1998. Laser-induced breakdown spectroscopy for polymer identification, Appl. Spectr. 52, 456.

[2] Wainner, R.T., R.S. Harmon, A.W. Miziolek, K.L. McNesby, and P.D. French, 2001. Analysis of environmental lead contamination: comparison of LIBS field and laboratory instruments, Spectrochim. Acta Part B, 56, 777.

[3] De Lucia Jr., F. C., R.S. Harmon, K.L. McNesby, R.J. Winkler Jr., and A.W. Miziolek, 2003. Laser-induced breakdown spectroscopy analysis of energetic materials, Appl. Opt., 42(30), 6148.

[4] Sturm, V., and R. Noll, 2003. Laser induced breakdown spectroscopy of gas mixtures of air, CO2, N2, and C3H8 for simultaneous C, H, O, and N measurement, Appl. Opt., 42(30), 6221.

[5] Noll, R., H. Bette, A. Brysch, M. Kraushaar, I. Mönch, L. Peter, and V. Sturm, 2001. Laser-induced breakdown spectrometry—applications for production control and quality assurance in the steel industry, Spectrochim. Acta Part B, 56, 637–649.

[6] Bohling, C., D. Scheel, K. Hohmann, W. Schade, M. Reuter, and G. Holl, 2006. Fiber-optic laser sensor for mine detection and verification, Appl. Opt., 45(16), 3817.

[7] Willer, U., M. Saraji, A. Khorsandi, P. Geiser, W. Schade, 2006. Near- and midinfrared laser monitoring of industrial processes, environment and security applications, Opt. Lasers Eng., 44, 699.

[8] Ch. Mann et al., 2004. Quantitative Detection of Nitric Oxide with Quantum Cascade Lasers, Proc. SPIE, 5365, 173.

[9] Willer, U., T. Blanke, and W. Schade, 2001. Difference frequency generation in AgGaS2: Sellmeier and temperature-dispersion equations, Appl. Opt., 40, 5439.

16. STAND-OFF DETECTION OF EXPLOSIVES OF SUICIDE BOMBERS BY MEANS OF OPEN-PATH FTIR SPECTROSCOPY

Konradin Weber[1], Christian Fischer[1], Günther van Haren[1], Horst Krause[2], Gudrun Bunte[2], Wenka Schweikert[2], and Thomas Härdle[2]
[1]*University of Applied Sciences (FHD), Düsseldorf, Germany*
[2]*Fraunhofer Institute for Chemical Technology (ICT), Pfinztal, Germany*

Abstract. In view of potential terroristic attacks on public areas like airports, metro stations, stadiums etc. it is important to detect suspect persons like suicide bombers before they can bring the explosive to detonation. This detection of the explosives must not be recognized by the suspect persons. It should deliver online results and should be performed in a stand-off way without direct contact to the suspect person (remote sensing). The open-path FTIR spectroscopy is a promising tool for that specific task. In this paper several applications are described and first measurement results for the online detection of TATP are presented. Moreover a short glance on further planned research items is given.

Key words: suicide bombers, open-path FTIR spectroscopy, explosives, remote sensing, stand-off detection, acetoneperoxide (TATP), DNMB, DNT, TNT, security, homeland security, airport, metro, stadium

16.1. Introduction

Explosives like acetoneperoxide (TATP) can be synthesized with relative low effort, although the handling of the substance is risky. There are informations that TATP was used for several terroristic attacks, e.g. for the terror attack on July 7th, 2005 in London and for the attack preparation for suicide bombers for the Christmas market of Strasbourg in the year 2000.

The basis for the stand-off detection of explosives by open-path FTIR spectroscopy is, that several explosives like TATP, DMNB-marked commercial explosives, DNT, TNT have the tendency of outgasing or sublimation and that the resultant vapour in the atmosphere can be measured by open-path FTIR spectroscopy, in principle, as there are strong absorption features in the infrared. The primary vapour pressure of these substances is varying between

H. Schubert and A. Rimski-Korsakov (eds.), Stand-off Detection of Suicide Bombers and Mobile Subjects, 135–143.

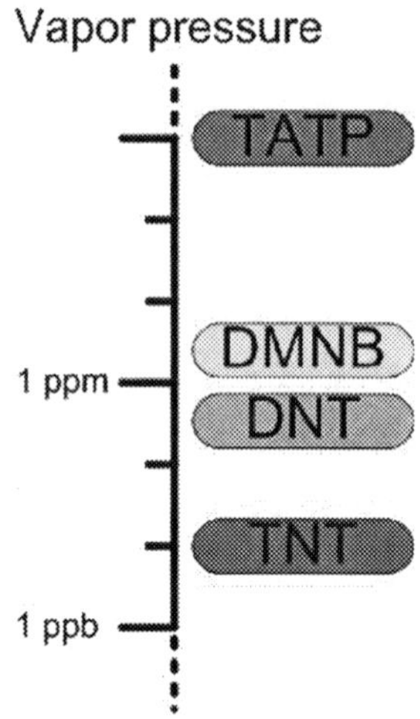

Figure 1. Characteristic vapour pressure for important explosives, from [1]

1000 ppm (TATP), around 1 ppm (DMNB, DNT) down to 10 ppb (TNT), see Figure 1. The concentrations to be measured in the open atmosphere of real world measurement sites are even lower. Moreover, the dispersion kinetics of the vapours of the explosives around the suicide bombers has to be taken into account, considering the sensitivity of the method. This has following background:

Suicide bombers tend to wear the improvised explosive devices around their body or e.g. in a back pack. Due to the thermal temperature of the body and the natural convection streams of the body heat around it as well as the air flow around the body or object during movement the suspected person will possess an explosive plume around its body. The produced plume concentration might be dependent on the type of clothes (e.g. thin or thick silk, wool, cotton, or leather) the suicide bomber is wearing. There might be an unhindered migration of the evaporated explosive through the clothes and consecutive passing to the surrounding air as well as a tendency of the targeted components to be "sticky" to the concealing material. Surely in practice the maximum concentration given by the equilibrium vapour pressures would not be reached (perhaps decreased concentrations by a factor of ten or hundred might be realistic).

16.2. The Open-Path FTIR Measurement Method

The open-path measurement systems work in a way, that they are transmitting infrared radiation into the open atmosphere on a measurement path of about ten

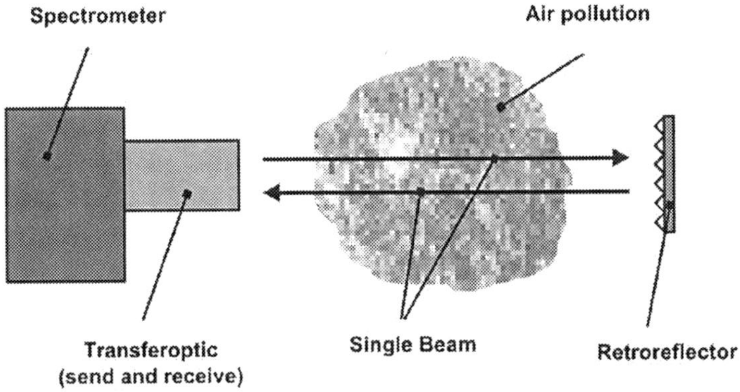

Figure 2. Schematic view of a monostatic open-path FTIR measurement system

to hundreds of meters towards a mirror or retroreflector, which reflects a part of the radiation back to the receiving unit and spectrometer. The concentration of the gases of interest is determined by the measurement of the absorption of the compounds within the measurement path using the FTIR spectrometer. In the so-called monostatic systems the IR source is implemented together with the spectrometer in one unit (see Figure 1). The so-called bistatic units use a separate IR source. More details of the measurement methods are described e.g. in [2–4].

16.3. Applications of Stand-Off Detection with Open-Path FTIR Spectroscopy

Applications for the stand off detection of explosives by their outgasing vapours with the open-path FTIR spectroscopy could be e.g.:

- airports (see e.g. schematic example in Figure 3);
- metro stations;
- public buildings;
- stadiums;
- large places (eventually, music festivals, sports events).

The measurements can be performed in a way, that measurement beams in the open atmosphere are established above the people along pathways, which most passengers have to pass, e.g. catwalks, escalators or waiting cues in front of counters. If a suspect person with outgasing explosives is hiding within the crowd of waiting people, it can be detected by the op-FTIR measurement beams.

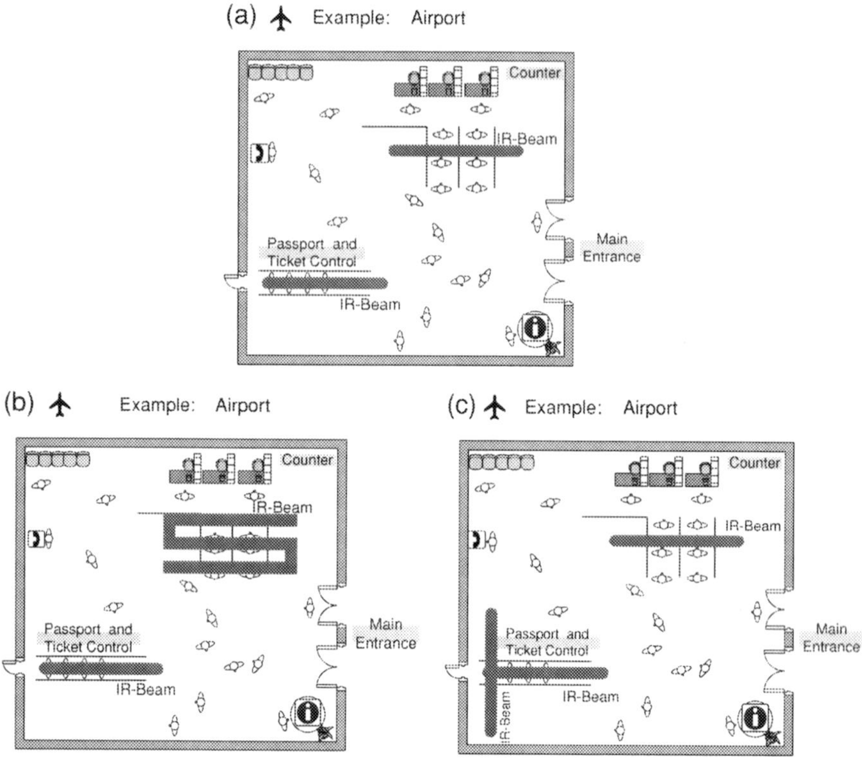

Figure 3. Example of application:Schematic views of possible examples of measurement beams of open-path FTIRsystems above waiting people within an airport area for the detection of vapours from explosives
a: Top view oft two beams of two open-path FTIR systems above waiting persons in front of counters and passport/ticket control
b: Folded measurement beam above waiting persons in front of ticket counter
c: crossed measurement beams (lefthandside) for pinpointing a single person

The advantages of the use of the open-path FTIR spectroscopy for this purpose are:

- The measurement beams are invisible;
- Large areas can be surveyed;
- The repetition rate of the measurement could be below a second, depending on the type of instrument and the resolution of the spectra;
- The technology is available, in principle, however, it has to be adapted to the measurement application;
- The measurement itself is not dangerous or harmful to the people;
- Interferences to other substances can be reduced by recording complete spectra.

Figure 3 demonstrates the possible application of this method in an airport: In Figure 3(a) the air above the waiting people in different areas is analyzed by the means of two measurement beams continuously. Figure 3(b) demonstrates, how the area, which is controlled, can be enlarged by folding the beam. By this way, the sensitivity of the measurement is enlarged as well. However, in this way it is only possible to detect, if any explosive like TATP is within this area of supervision. It is unlike to pinpoint a single person with this method. This can be done in a configuration shown in Figure 3(c). Here an additional instrument with a measurement beam rectangular to the primary beam is used. By crossing the measurement beams the pinpointing of single persons carrying the explosives (suicide bombers) should be possible.

Similar applications are possible in metro stations, stadiums and public areas.

16.4. First Measurement Results

Figure 4 shows the measurement setup for first TATP test measurements performed jointly by the FHD and ICT. The measurements were performed with a FTIR spectrometer, which is normally used by the FHD for open-path measurements. For the test measurement this FTIR system was coupled with a multireflection cell, realizing a measurement path of 40 m. In this way realistic conditions for the measurement path length could be achieved, while

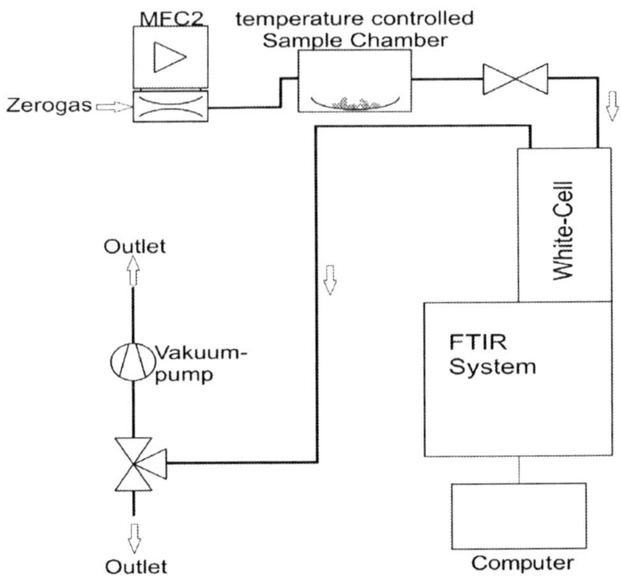

Figure 4. Schematic view of measurement setup

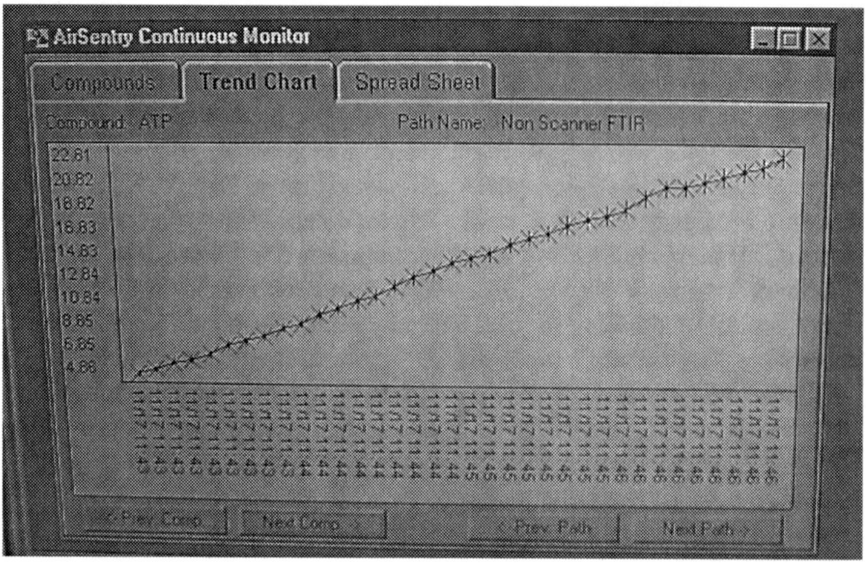

Figure 5. Screenshot of measured increasing TATP concentrations

simultaneously controlled atmospheric conditions could be guarantied for the test measurement.

The measurements of acetoneperoxide (TATP) were performed in following way: The TATP was transferred to a temperature controlled sample chamber, which was slowly heated with a constant heating rate. In parallel to the constant heating zero gas was flushed through the sample chamber with a constant flow rate, which was adjusted by a mass flow controller. By this means air is fed through the multireflection cell, which is enriched with a steadily increasing concentration of TATP vapour.

During the measurements it was investigated, if the FTIR system would be able to detect continuously the TATP vapour in the air. For that analysis purpose a TATP reference spectrum was prerecorded.

Figure 5 shows a screenshot of the FTIR computer during the measurement. The screenshot displays the increasing TATP concentrations on the 40 m measurement path in the air of the multireflection cell. This is due to the outgasing effect of the TATP during heating in the temperature controlled heating chamber. In Figure 5 it can be seen clearly that the FTIR system was able to monitor the increasing TATP concentrations instantaneously and continuously. Figure 6 shows spectra, recorded by the FTIR spectrometer at the FHD, of the air with (lower line) and without the TATP vapour (upper line). As it is demonstrated clearly in Figure 6, the TATP vapour causes lower measured intensity in several regions of the spectrum due to absorption (for instance in the region around 1000 wavenumbers). This becomes evident also by comparing the measured spectrum with the prerecorded reference spectrum of

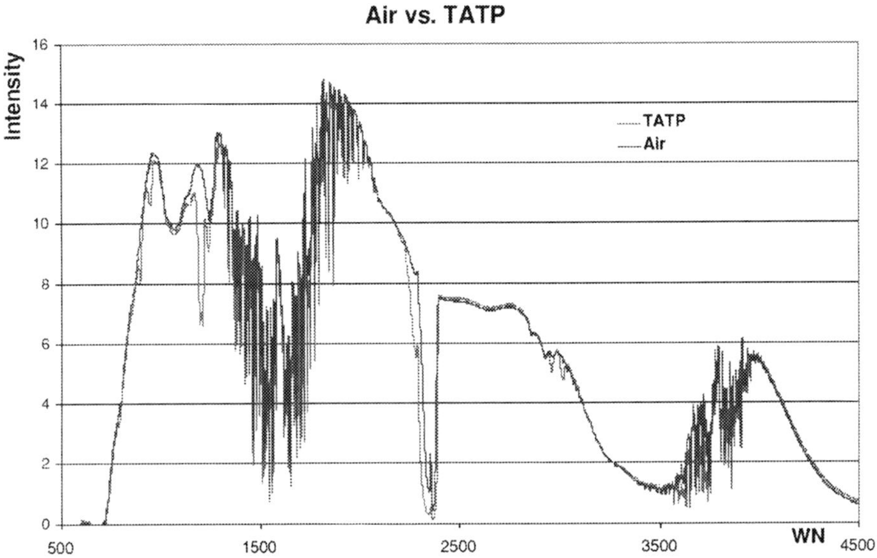

Figure 6. Measured spectra of air with TATP (lower line) and without TATP (upper line)

TATP (Figure 7, taken by ICT). Therefore, by these measurements it could be proven, that explosives like TATP can be measured continuously by the open-path FTIR spectroscopy.

Moreover, reference spectra were recorded for DMNB, DNT and TNT as well, so measurements should be also possible for those compounds.

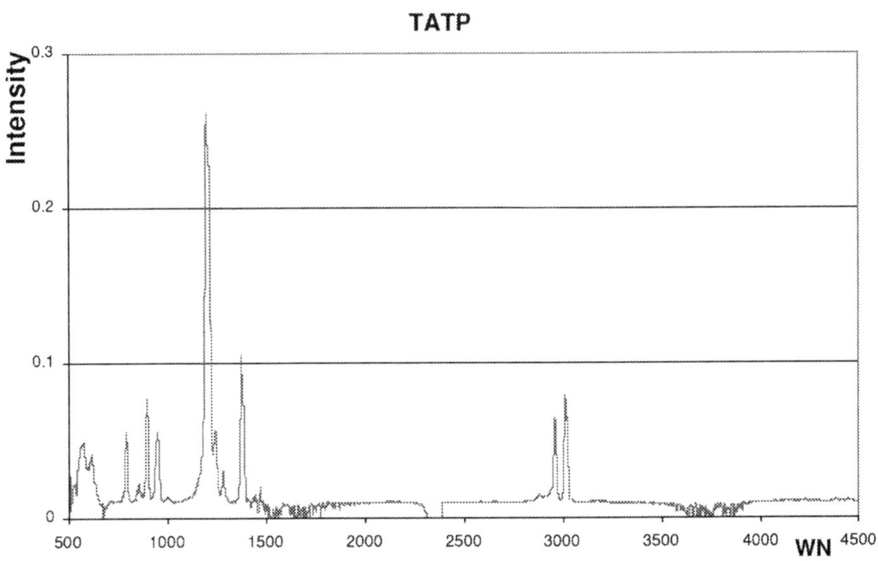

Figure 7. TATP reference spectrum

16.5. Further Research Items

In the described measurements the University of Applied Sciences Düsseldorf (FHD) and the Fraunhofer Institute for Chemical Technology (ICT) clearly demonstrated the ability of the open-path FTIR spectroscopy to detect explosives like TATP by their outgasing vapours in the atmosphere. Based on these measurement results, a joint research project is planned by the FHD and the ICT.

The University of Applied Sciences Düsseldorf (FHD) has already utilized and developed the open-path spectroscopy for several applications [5–9]. The Fraunhofer Institute for Chemical Technology (ICT) in Pfinztal, Germany, has the expertise for the handling of explosives and analysis of the explosives with point analysers [10, 11]. The expertise of both institutions will be used in the joint project. Research items will addressed such as

- Investigation and optimisation of the detection limit of the open-path FTIR method depending on the response time and resolution of the analyzer.
- Investigations of interferences.
- Optimization of software for this specific measurement task.
- Investigation of the dispersion kinetics of the vapours of the explosives around suicide bombers, by sampling of the vapour plume around dummies with appropriate point monitoring techniques.

16.6. Conclusion

It could be demonstrated successfully by the University of Applied Sciences Düsseldorf (FHD) and the Fraunhofer Institute for Chemical Technology, that the open-path FTIR spectroscopy is a promising tool for the detection of explosives like TATP, which are used by suicide bombers for terroristic attacks. This detection technique could be used for the surveillance of public areas like airports, metro stations or stadiums etc. Further research on this topic is planned. This would be an important step for public security.

Acknowledgements

The equipment for this work was partly funded by the Ministry of Innovation, Science, Research and Technology of the State of North Rhine-Westphalia, Germany, and the Federal Ministry of Education and Research of Germany.

References

[1] National Research Council, Existing and Potential Standoff Explosives Detection Techniques, National Academic Press, Washington, DC, www.nap.edu.

[2] US Environmental Protection Agency, 2001. TO 16: Standard Guide for Open Path Fourier Transform Infrared (OP/FT IR) Monitoring of Gases and Vapors in Air, Research Triangle Park, USA.

[3] VDI-Guideline 4211, 2000. Remote Sensing—Atmospheric Measurements Near Ground with FTIR Spectroscopy—Measurement of Gaseous Emissions and Immissions, Beuth Verlag, Berlin, 76 p.

[4] Weber, K., A. Ropertz, T. Lamp, and G. van Haren, 1998. FTIR-Measurements for Air Pollutants on Long Measurement Paths in the Open Atmosphere—Basic Principle and Quality Assurance Aspects, Proceedings of the 11th World Clean Air Congress, Durban, South Africa, 1, IUAPPA NACA, Parklands, South Africa, No. 13B-3.

[5] Weber, K., T. Lamp, K. Schulte, J. Weidemann, G. van Haren, A. Gärtner, and G. Bröker, 1996. Open-Path FTIR-Measurements of Methane as a Tracer Gas for Waste Site Emissions at a Waste Site in Germany, Proceedings of the 89th Annual Meeting & Exhibition of the Air & Waste Management Association, A&WMA Pittsburgh, 11p.

[6] Lamp, T., K. Weber, A. Ropertz, and G. van Haren, 1998. Applications of Open Path FTIR Measurement Technique for Measurement of Air Pollutants at Industrial Sites, Waste Sites and for Measurements of Greenhouse Gases, Proceedings of the 11th World Clean Air Congress, Durban, South Africa, 1, IUAPPA NACA, Parklands, South Africa, No. 4D-4.

[7] Kuttler, W., T., Lamp, and K., Weber, 2002. Summer air quality over an artificial lake, Atmospheric Environment, 36 (39–40), pp. 5927–5936.

[8] Weber, K., Ropertz, A., Fischer, Ch., and G. van Haren, 2003. Application and Use of Optical Remote Measurement Systems (OP-FTIR and UV-DOAS) during a One-Year Measurement Campaign for Ambient Air Quality at an Urban Area in Germany, Proceedings of the 96th Annual Meeting of the Air & Waste Management Association, A&WMA Pittsburgh, PA.

[9] Weber, K., C., Fischer, G. van Haren, A. Ropertz, A. Chaudhry, 2005. Measurements of fugitive emissions of a biogas production plant in Germany with op-FTIR and conventional measurement systems, Proceedings of the 98th Annual Meeting & Exhibition of the Air & Waste Management Association, A&WMA Pittsburgh, PA.

[10] Bunte, G., J. Deimling, H. Krause, 2004. Evaluation of the SPME-enrichment for the fast and automatic detection of explosives in aqueous samples by GC-MSD, Proceedings of the 8th International Symposium on the Analysis and Detection of Explosives, edited by Donna Garbutt et al., Ottawa, Canada, September, 2004.

[11] Bunte, G., J. Hürttlen, G. Unkelbach, K. Niedergall, and H. Krause, 2005. Gas phase detection of TNT with imprinted polymers, 36th International Annual Conference of ICT, Energetic Materials—Performance and Safety, June 28–July 1, 2005, Karlsruhe, Germany.

17. ^{14}N NQR DETECTION OF EXPLOSIVES

C. Fermon[1], J.F. Jacquinot, M. Pannetier-Lecoeur, J. Scola, and A.L. Waillang
Commissariat à l'Energie Atomique, DSM/DRECAM/SPEC 91191 Gif-sur Yvette Cedex, France

Abstract. Nuclear Quadrupolar resonance allows selective detection of solid explosives containing Nitrogen. This note presents results on TNT, HMX and RDX by a surface coil measurement.

Key words: NQR, explosive detection

17.1. Introduction

NQR is a solid state spectroscopic technique that applies to nuclei of spin $I \geq 1$ in a non-cubic environment. The interaction between the nuclei, characterised by their electric quadrupole moment, and the environment characterised by the components of the electric field gradient at the nucleus gives rise to discrete energy levels between which resonant transitions can be induced by a radiofrequency magnetic field. The resonance frequencies are compound specific and allow detection of various kinds of objects (bombs, drugs, ...) on the basis of their chemical composition, which is an important advantage over other methods that detect extrinsic properties such as metallic mine detectors. The technique requires that the substance is in a solid state and that the environment of the nuclei of interest is non-cubic, otherwise the electric field gradient is zero because of symmetry consideration in the cubic case, or because of motion averaging in the case of a liquid substance.

Most solid explosives contain an important proportion of nitrogen which allows their detection by NQR. The resonance frequencies lie in the 500 kHz–5 MHz range.

In its simplest variant the NQR detection experiment consists of an excitation period which disturbs the nuclei from their equilibrium state, followed by a detection period during which the oscillations of the nuclei at the resonance frequency are monitored. The excitation is a radiofrequency pulse

[1] To whom correspondence should be addressed.

H. Schubert and A. Rimski-Korsakov (eds.), Stand-off Detection of Suicide Bombers and Mobile Subjects, 145–150.

of appropriate amplitude and length. In the classical scheme of detection, the oscillations of the nuclear magnetisation at the resonant frequency after the pulse are detected through the voltage they induce in a coil by the Faraday effect. Usually the same coil is used for excitation and detection.

The important parameters that govern the efficiency of detection for a given nucleus are its concentration, the resonance frequency and the values of several characteristic times, listed non exhaustively below, that characterise the return of nuclei toward equilibrium after the initial disturbance:

T_1 is the time of thermal equilibration between the various energy levels, T_2* is the time of decrease of the magnetisation created by initial pulse, including extrinsic factors such as inhomogeneities (temperature, resonance frequency ...) inside the sample.

$T_2 > T_2$* is the intrinsic time of loss of magnetisation. It does not necessarily involve transfer of population between the energy levels and one has $T_2 < T_1$.

For an optimum detection of a substance, the knowledge of the relaxation times is crucial. It allows optimising the simple detection sequence described above, but it also allows improving upon this simple sequence. Depending on the NQR characteristics, various elaborate sequences have been proposed: the point common to these sequences is that they use multipulse excitation which permit to refocus the magnetisation and to increase the effective observation time above the value T_2* which pertains to the simple one pulse excitation sequence.

From the historical point of view the detection of explosives by NQR has been considered very early in several countries. In USA extensive work has been conducted by the team of Garroway with a first patent in 1993 [1]; in Great Britain one can mention the works of Smith and al. with several patents starting from 1989 [2], and a demonstration of the possibility to detect ammonium nitrate inside a vehicle [3]; in Russia NQR studies are associated with the name of Grechishkin [4]. In Australia, a company (QRScience) [5] is developing a commercial explosives detector for airport.

However, NQR systems have until now not really been implemented in real conditions due to the fact that the sensitivity is still rather low, especially for TNT which presents a low density of Nitrogen and long relaxation times. The limitation in sensitivity is given by the thermal noise of the detecting coil. For that reason, we are exploring alternative solutions based on cooled sensors.

17.2. Experimental Setup

Practically, a NQR experiment consists in emitting RF pulses into the sample and then collecting its nuclear magnetization response. We have developed

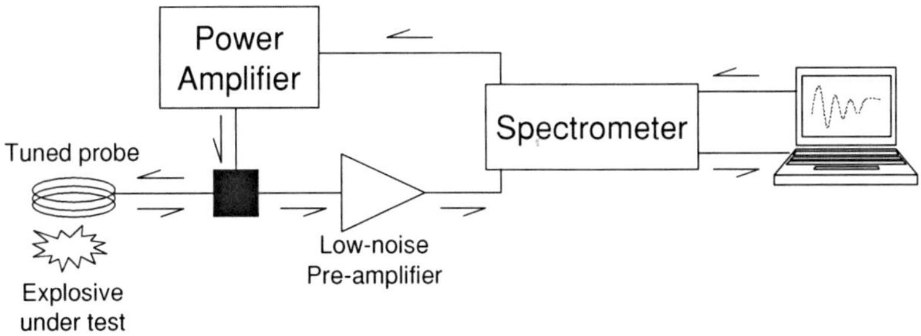

Figure 1. Experimental setup. A surface coil is connected to a home-made spectrometer controlled by a PC. The probe and the preamplifier are separated from the remaining part by several meters

a high performance spectrometer which covers a frequency range between 400 kHz to 10 MHz, 0.1° phase accuracy, 100 db amplitude control and 10 ns pulse resolution. A double frequency change associated to very low noise preamplifiers in the reception part insures an optimized signal-to-noise ratio of the electronics. Then the effective noise is given mainly by the thermal noise of the resonant coil (see Figure 1).

Our system can work simultaneously with an optimized resonant surface coil and a cooled sensor. The coils used in the experiments have a solenoid shape (radius = 4.5 cm, length = 1 cm), and are designed to optimize the homogeneity of the RF field in the explosive. The working distance is comprised between 5 and 10 cm. The cooled sensor has about the same size with an additional 3 cm distance from the explosives.

The pulse power was at maximum 200 W with pulse lengths between 5 μs and 100 μs depending on the system.

17.3. Experimental Results

We have performed experiments on various explosives and we present here results obtained on HMX, RDX and TNT. The measured quantities were 100 g and the temperature was 5 °C.

RDX has a rather short T_1 of ~18 ms at room temperature and then can be easily and rapidly detected in a single pulse experiment. Figure 2 shows the spectrum around the 3.4 MHz line. Acquisition time was 10 s. This time can be easily shortened by using a SORC sequence [6].

HMX presented a very short T_2 relaxation time. For that reason, we used a spin echo sequence for its detection. Figure 3 shows the time signal for HMX.

TNT is harder to detect because the relaxation time is longer (6 s), the resonance frequency is lower (0.8 MHz) and the N density is smaller.

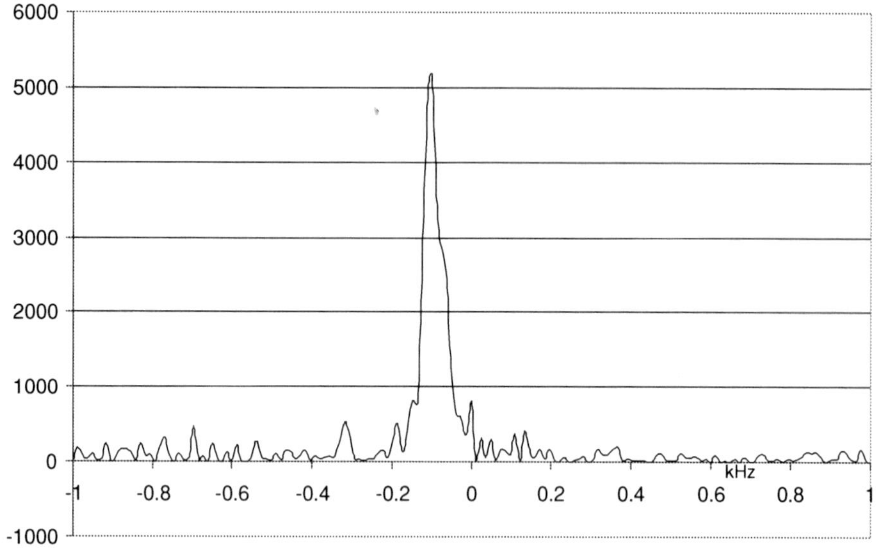

Figure 2. RDX signal detected at 3.4 MHz with a single pulse sequence. The signal is given in arbitrary units as function of frequency offset in kHz

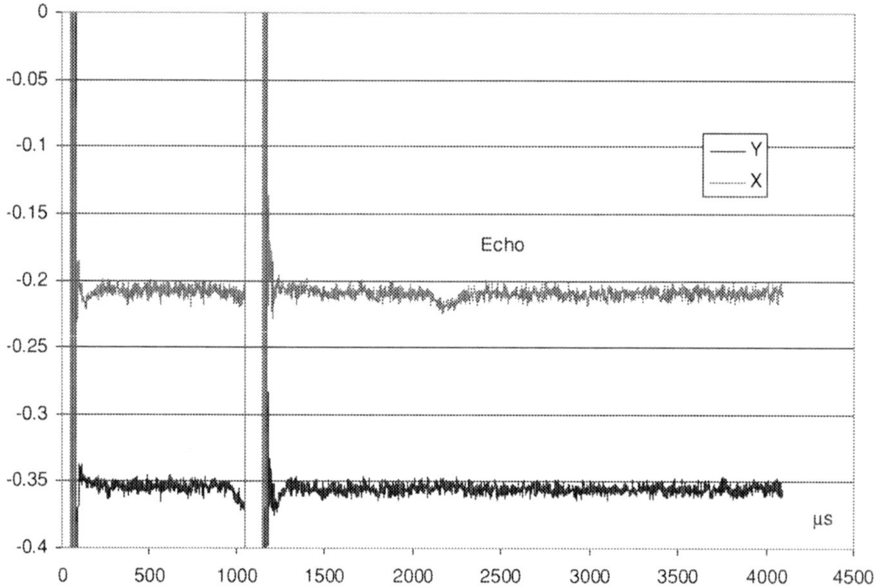

Figure 3. HMX spin echo signal. We can see the first pulse, the second RF pulse and the echo. The working frequency is 3.6 MHz. The echo can only be seen in phase with the RF frequency. The signal is given in arbitrary units as function of time in µs

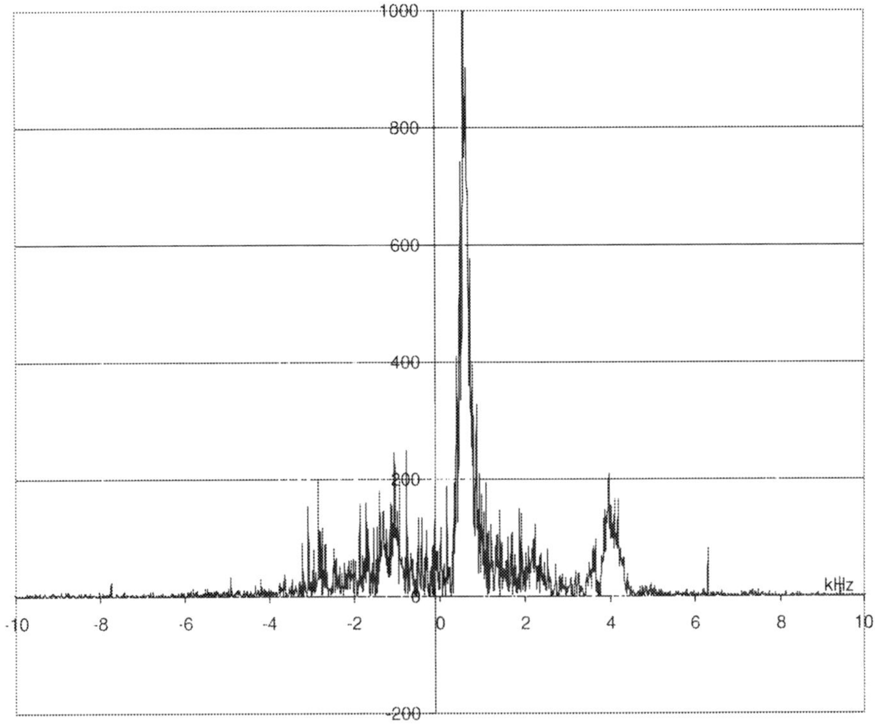

Figure 4. TNT spectrum observed with a single pulse sequence at 0.840 MHz. The signal is given in arbitrary units as function of frequency offset in kHz. No post data treatment has been performed

However we detected it rather easily, Figure 4 shows the signal observed at 0.840 MHz.

17.4. Perspectives

In the experiment performed, the filling factor was rather high with an explosive placed at 5 cm of a surface coil. The sensitivity of our spectrometer is enough good to detect easily less than 100 g of TNT or other solid explosives.

In a real experiment like detection in luggage, the filling factor will be less favorable (ou lower ou smaller) and sensitivity problems will appear. For that reason, we are investigating alternative solution for RF detection based on cryogenic sensors coupled to flux-flux transformers which should give one to two orders of magnitude more sensitivity than present technology.

References

[1] Buess, M.L., A.N. Garroway, and J.B. Miller, 1993. Detection of explosives by nuclear quadrupole resonance, US Patent no. 5206592 (April).

[2] Smith, J.A.S., 1989. Method and apparatus for detecting certain compounds, US Patent no. 4887034 (December).

[3] Barras, J., M.J. Gaskell, N. Hunt, R.I. Jenkinson, K.R. Mann, D.A.G. Pedder, G.N. Shilstoneand, and J.A.S. Smith, 2004. Detection of ammonium nitrate inside vehicles by nuclear quadrupole resonance, Appl. Magn. Reson., 25 (3–4), 411–437.

[4] Grechishkin, V.S., 1992. NQR device for detecting plastic explosives, mines and drugs, Appl. Phys. A, 55, 505–507.

[5] http://www.qrsciences.com.

[6] Klainer, S.M., T.B. Hirschfeld, and R.A. Marino, 1982. In *Fourier, Hadamard and Hilbert Transform in Chemistry*, Plenum, New York, p. 147.

18. STAND-OFF EXPLOSIVES DETECTION USING TERAHERTZ TECHNOLOGY

Michael C. Kemp,[1] Colin Baker,[2] and Ian Gregory[2]
[1]*Iconal Technology Ltd., St. John Innovation Centre, Cambridge CB4, 0WS, United Kingdom*
[2]*TeraView Ltd, Platinum Building, St John's Innovation Park, Cambridge, CB4 0WS, United Kingdom*

Abstract. Terahertz imaging and spectroscopy has been shown to have the potential to use very low levels of this non-ionising radiation to detect and identify objects hidden under clothing at stand-off distances. In this paper we discuss some of the important factors involved in developing systems for the security industry, and describe our recent work on the development of a prototype terahertz stand-off detection system. Using this system we demonstrate the spectroscopic detection of concealed explosives at a stand-off distance of 1 m, both real time, in reflection, and under normal atmospheric conditions. We believe the technique has the potential for further development towards a practical system for the detection of suicide bombers and mobile subjects.

18.1. Introduction

The terahertz (THz) region of the electromagnetic spectrum is typically considered to occupy 300 GHz to 10 THz, bridging the gap between millimeter waves and the infrared. Recent developments have started to address the previous lack of sources and sensitive detectors in this range [1, 2]. These advances have been accompanied by much interest in possible applications of the technology, just as has happened in the past as other parts of the spectrum have become accessible. Interest in terahertz technology has focussed on the security [3, 4], pharmaceutical [5], non-destructive testing (NDT) [6], and medical industries [7, 8].

There are unique properties of THz radiation that make it a potentially powerful technique in security screening. Firstly, THz radiation penetrates many everyday physical barriers such as typical clothing and packing materials with modest attenuation [3]. Secondly, many chemical substances and explosive materials exhibit characteristic spectral responses at THz frequencies that can be used to identify these threat objects [4]. Additionally, being at low

H. Schubert and A. Rimski-Korsakov (eds.), Stand-off Detection of Suicide Bombers and Mobile Subjects, 151–165.

photon energy, sub-millimeter wavelengths, THz radiation is non-ionising, in contrast to penetrating radiation using higher energies. Terahertz techniques, therefore, combine safe-to-use high-resolution imaging, and the identification through spectroscopy of threat materials—even when hidden in packages or under clothes.

Although they are eventually absorbed by water vapour in the atmosphere, terahertz waves will propagate modest distances through normal atmospheric air, so that terahertz can be considered for stand-off detection over distances of several tens of metres.

THz imaging systems can be passive, simply detecting the THz part of the thermal black-body radiation given off by objects. Alternatively, active schemes can be implemented whereby the object being imaged is illuminated by a THz source. Whilst passive systems may be effective for 2D low frequency (e.g. 100 GHz) millimeter wave imaging, chemical/structural analysis of suspect objects is restricted to higher frequency techniques, since there are virtually no spectroscopic features in solids below 500 GHz. Active techniques, such as terahertz pulsed imaging (TPI), and terahertz continuous-wave (TCW) imaging can be several times more sensitive. Pulsed and multi-frequency techniques enable 3D imaging, much like radar by using time of flight analysis providing increased contrast and discrimination.

Note that, as conceived here, the detection concept is to shine terahertz radiation, through clothing or bags carried by a suspect and spectroscopically to detect hidden explosives using reflections from the *solid* surfaces of the explosive material. Some explosives do have non-negligible vapour pressure (although many have very low vapour pressure) and whilst terahertz radiation will excite rotational modes in any emitted gas or vapour, these signals will be far too small to detect in any practical stand-off system. Terahertz stand-off detection is thus a bulk rather than trace detection technique although, as we shall demonstrate, it is capable of detecting small (and thin) macroscopic samples.

In this paper, we introduce terahertz technology, discuss some of the important factors involved in addressing practical systems for stand-off detection, and describe a laboratory proof-of-principle demonstrator we have constructed—a stand-off detection system, aimed at detecting threats such as explosives using spectroscopy at a distance.

18.2. Terahertz Systems

Established experimental THz systems are largely based on photoconductive switches, which rely on the production of few-cycle THz pulses using an ultrafast (femtosecond) laser to excite a biased photoconductive antenna. This causes the production of charge carriers (electrons and holes) which are accelerated by the bias voltage and these accelerated charges radiate. The technique is inherently broadband, with the emitted power distributed over several THz (typically 0.1–4 THz). Pulsed THz emission in photoconductive antennas is produced when the current density, j, of a biased semiconductor is modulated on subpicosecond timescales $E_{\mathrm{THz}} \propto \mathrm{d}j/\mathrm{d}t$ [9]. The change in current density, and hence photocurrent, arises from two processes: the rapid change of the carrier density via femtosecond laser illumination, and the acceleration of photogenerated carriers under an external electric field [10].

Coherent detection of the incident THz radiation can be performed in a similar photoconductive antenna circuit [11–14]. By gating the photoconductive gap with a femtosecond pulse synchronised to the THz emission, a DC signal that is proportional to the THz electric field may be measured. Further, by varying the optical path length to the receiver, the entire THz time domain can be sampled. In this way, both the amplitude and phase of the incident THz wave can be obtained, and a dynamic range of over 60 dB demonstrated using time-gated detection [15].

In a typical pulsed THz system (see Figure 1), the beam from the ultrafast femtosecond laser is split by a beamsplitter into two components, a pump beam

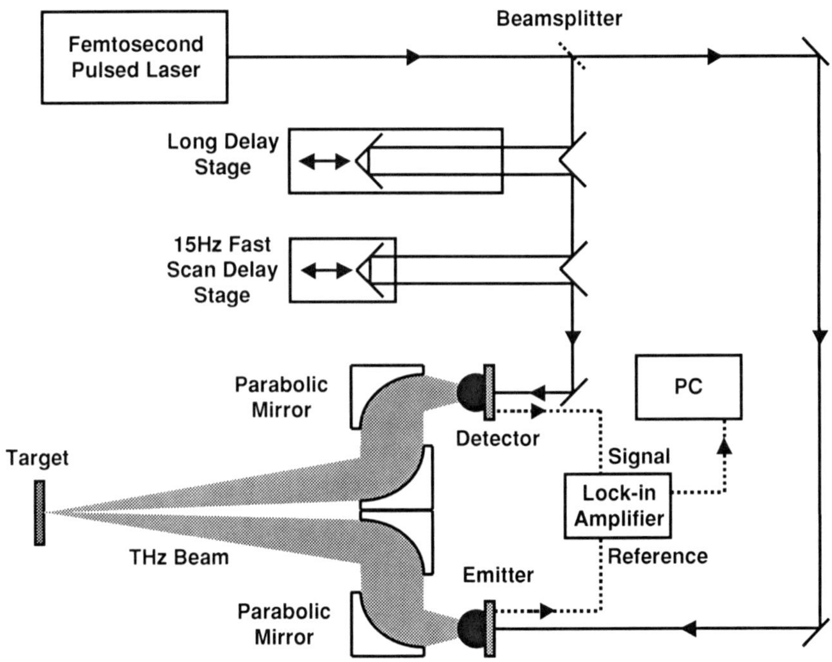

Figure 1. Schematic of a THz photoconductive system. In this example, specifically designed for stand-off explosives detection, the THz beam is manipulated for reflection spectroscopy of a target material

and probe beam, used to illuminate the emitter and receiver respectively. A motorised delay stage is then incorporated into the probe beam to vary the difference in optical delay around zero between the incoming THz pulse and the probe laser pulse at the detector.

A rapid scanning delay line is often utilised since it allows both the delay position and the lock-in output to be digitized and re-interpolated to obtain the THz field as a function of optical delay in real time. The frequency Fourier transform can then also be displayed in real time on a computer display. Typically, the output from the THz emitter is coupled from the rear surface of the device using a high-resistivity silicon lens, which, in combination with off-axis parabolic mirrors/lenses, allows the THz beam to be manipulated as required. Additionally, to couple the THz radiation from free-space into the receiver, a second silicon lens in contact with the rear of the receiver chip is used.

The THz emitter and detector are manufactured from SI and LT GaAs materials respectively, using, for the detectors, recently developed materials processing techniques [16]. A bias of 50 V was applied across the emitter electrodes, and the gated output signal from the receiver was fed to a lock-in amplifier.

18.3. Technical Issues

If terahertz techniques are to be suitable for stand-off explosives detection of hidden explosives, as in the detection of suicide bombers, a number of conditions need to be established:

- the target explosives must have characteristic spectroscopic features which allow them to be identified;
- these features must be distinct from harmless, potential confusion materials;
- it must be possible to carry out measurements in a reflection geometry;
- barrier materials such as clothing and baggage materials must be sufficiently transparent to THz and must not introduce spectroscopic features of their own;
- propagation through the atmosphere must be sufficient;
- system performance must be good enough to permit real time (or very near real time) operation.

There are also detection quality and concept-of-operations issues which are not discussed here.

All these issues have been investigated in a programme of work carried out by TeraView over the last three years and some of the results are described below.

18.4. Terahertz Spectroscopy of Explosives and Other Materials

To validate that THz technology is capable of detecting explosives, we first measured the spectrum of various common explosives in transmission. This was performed using our commercially available transmission spectrometer, TPIspectra1000. In transmission the THz field is modified by dispersion and the absorption of the media under examination. The ratio of the electric field strength before $E_r(\omega)$ and after transmission $E_s(\omega)$ is given by

$$\frac{E_s(\omega)}{E_r(\omega)} = t_{n(\omega)} \exp\left(-\frac{\alpha(\omega)d}{2} + \frac{(n(\omega)-1)\omega d}{c}\right)$$

$$t_{n(\omega)} = \frac{4n(\omega)}{(n(\omega)+1)^2}$$

where d is the thickness of the sample, ω the frequency of the radiation, c the speed of light, and $t_n(\omega)$ is the reflection loss at the sample surface as defined above. Hence, in transmission spectroscopy both the refractive index $n(\omega)$ and the absorption coefficient $\alpha(\omega)$ can be determined from the measured amplitude and phase information by

$$\alpha(\omega) = \frac{2}{d}\left[-\Re\left\{\log\frac{E_s(\omega)}{E_r(\omega)}\right\} + \log\frac{4n(\omega)}{(n(\omega)+1)^2}\right] \quad (1)$$

$$n(\omega) = 1 + \frac{c}{\omega d}\Im\left\{\log\frac{E_s(\omega)}{E_r(\omega)}\right\}. \quad (2)$$

We first published a number of spectra of common energetic compounds (RDX, PETN, HMX, TNT) and commercial explosives based on these compounds (PE4, Semtex-H) in 2003 [3]. These initial results are shown in Figure 2 and have subsequently been validated and extended both by ongoing work in our group and by other independent groups.

In Figure 3 we show some recent measured absorption spectra and refractive index of RDX, and PETN. We note from these measurements that there are clear and unique spectral features corresponding to each of the explosives. These results agree well with previous reports [3], but with extended spectral coverage (5–120 cm^{-1}), and superior spectral resolution (1 cm^{-1}).

In Figure 4 we show the absorption spectrum (solid line) and refractive index (dashed line) of Semtex-H. The features appearing in the absorption spectrum of Semtex-H can be understood from the constituent explosives. Semtex-H consists of approximately equal amounts of RDX and PETN, with the remainder comprising of poly(butadiene-styrene) and oil. It can be seen that the absorption spectrum of Semtex-H is the sum of the spectra of its active constituents, shown as the dotted line.

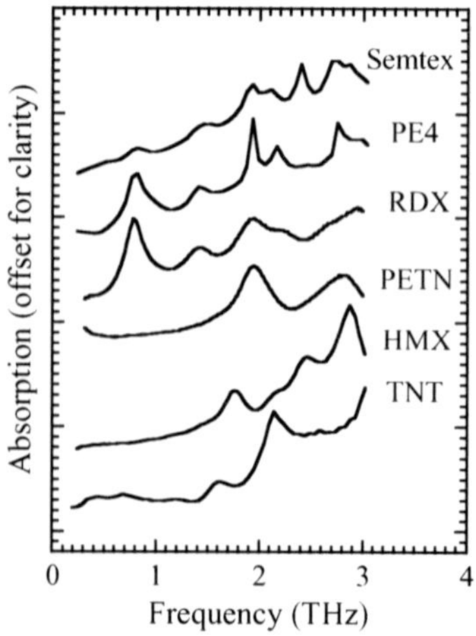

Figure 2. Terahertz transmission spectra of the raw explosive materials TNT, HMX, PETN and RDX together with the spectra of the compound explosives PE4 and Semtex-H

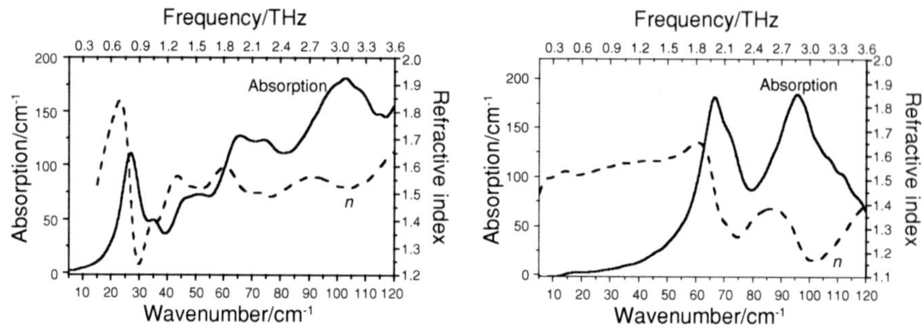

Figure 3. The THz absorption spectra (solid line) and refractive index (dashed line) of (a) RDX and (b) PETN

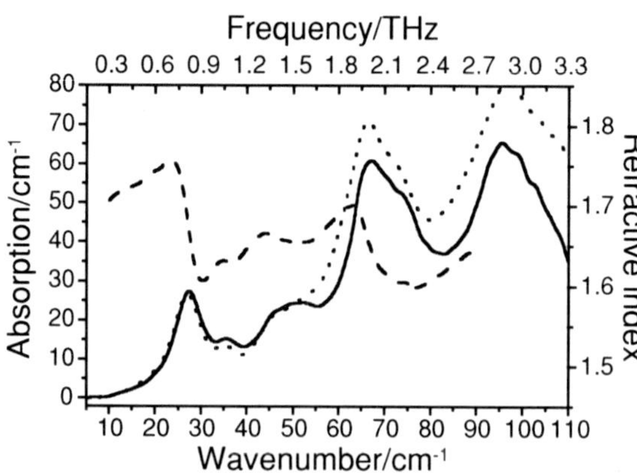

Figure 4. The THz absorption spectrum (solid line) and refractive index (dashed line) of Semtex-H. The dotted line shows the combined absorption spectrum of RDX and PETN

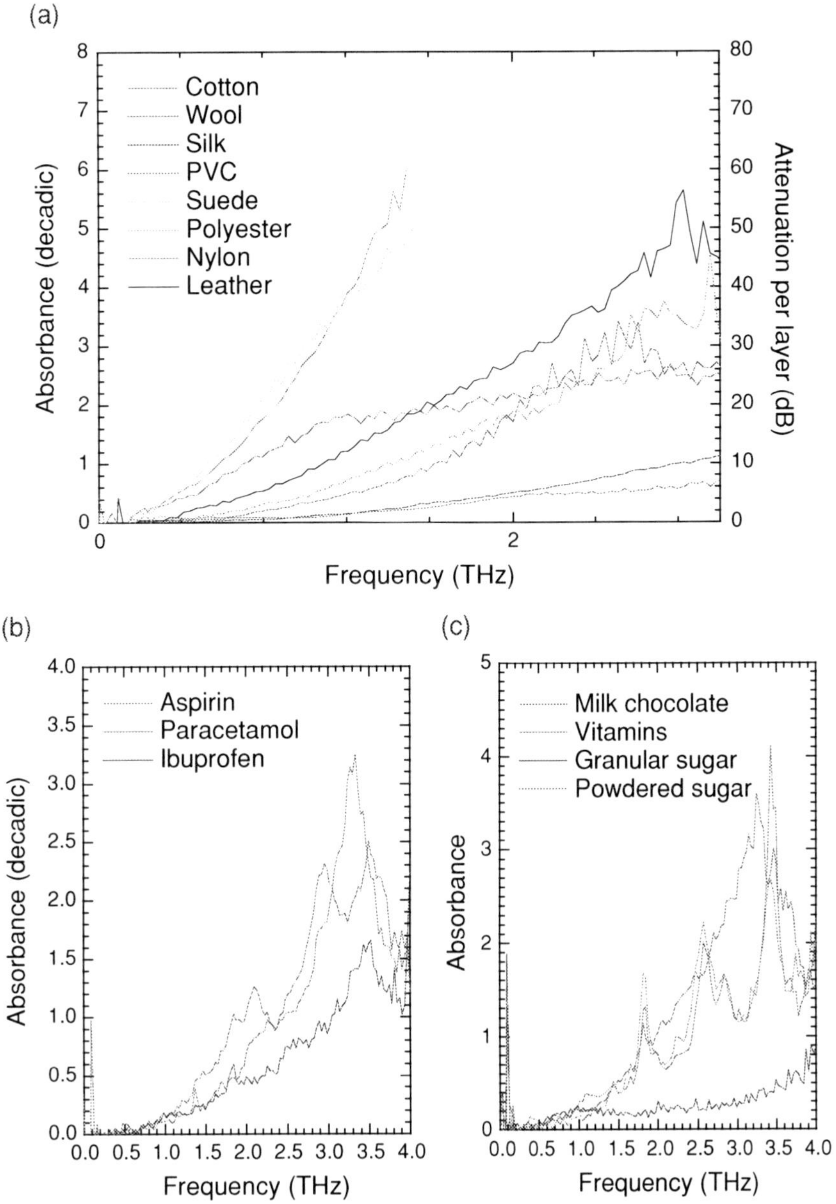

Figure 5. (a) Clothing absorption measurements. Absorption spectra from (b) household drugs, and (c) food items and supplements

We have carried out several studies and measurement programmes to characterise common barrier materials and potential confusion materials [4]. Indicative results are shown in Figure 5. Most barrier materials such as different types of cloth, paper, cardboard, plastics are semi-transparent to terahertz with an absorption which rises smoothly with frequency. Confusion materials, such as foodstuffs, confectionery, cosmetics, pharmaceuticals either have featureless spectra or may have spectroscopic features in the THz range. However, having examined a large number of substances, we find spectra are distinct and we have not observed significant confusion between explosives and other materials.

Despite the high sensitivity and dynamic range of THz pulsed spectroscopy (TPS) techniques any practical implementation of a people screening system will need to work in reflection rather than transmission geometry. From the refractive index, as determined from transmission spectroscopy, the reflectance can be determined. The reflectance, defined as the frequency dependent ratio of the reflected intensity to the incident intensity, is related approximately to the refractive index by

$$R(\omega) \approx \left(\frac{n(\omega) - 1}{n(\omega) + 1}\right)^2.$$

Transmission spectroscopy measurements can therefore be used to calculate reflection spectra. Figure 6 shows the expected reflectance of Semtex-H as determined from the above equation.

The contrast in THz reflection spectra are significantly less than in transmission/absorption measurements which makes detection more challenging, but distinct spectral features are nonetheless present. We have carried out theoretical studies into stand-off detection using reflection measurements including the effects of propagation geometry, clothing barriers and atmospheric absorption. These indicate that it is possible to build a stand-off detection system using THz. In the next section, we describe the construction of a laboratory proof-of-principle demonstration to validate this conclusion.

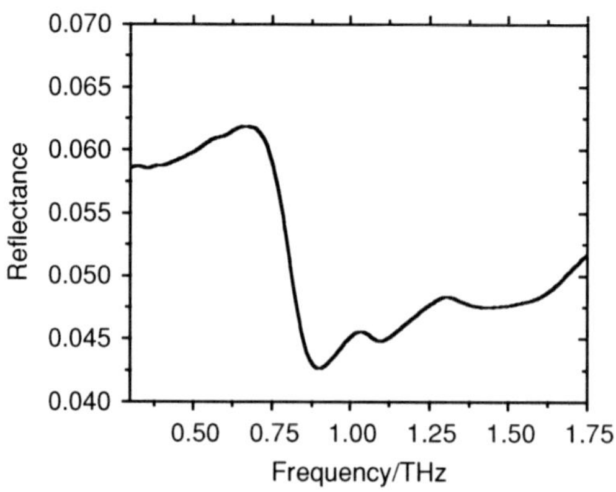

Figure 6. The calculated reflection spectrum of Semtex-H, as determined from transmission spectroscopy measurements

18.5. Proof-of-Principle Demonstration

In order to explore stand-off detection using THz, we constructed a laboratory proof-of-principle demonstrator with an initial target distance of 1 m. Using this system we were able to demonstrate for the first time the spectroscopic detection of concealed explosives at a stand-off distance of 1 m, real time, in reflection, and under normal atmospheric conditions.

A schematic of the pulsed THz photoconductive system used is shown in Figure 1. In our setup, a Coherent Vitesse laser was used to generate pulses of average power 800 mW, at 800 nm centre wavelength, 80 MHz repetition rate, and with a bandwidth limited pulse duration of 80 fs. As described in Section 2, the output of the ultrafast laser was split into two components, a pump beam and probe beam, used to illuminate the emitter and receiver respectively. This was achieved by using a high-energy beamsplitter, with 80% of the beam being reflected into the probe beam, and 20% transmitted as the pump beam. A rapid scanning delay line oscillating at 15 Hz, and a 1000 mm delay stage were incorporated into the probe beam. The optical laser powers at the emitter and receiver were controlled by using neutral density filter wheels in both the pump and probe beams. In this arrangement, the pump and probe laser powers incident on the photoconductive devices were measured to be 5 mW, and 20 mW respectively.

The optical scheme for the THz radiation was as follows: the generated THz radiation from the photoconductive emitter was coupled from the rear surface of the device using a high-resistivity silicon hyperhemispherical lens, which, in combination with an $f/1$ off-axis gold parabolic mirror, collimated the THz beam. This was then focused and directed onto a target object at a stand-off distance of 1 m by a second off-axis parabolic reflector. The reflected THz beam from the target object then followed an identical reverse scheme, using two parabolic mirrors and a hyperhemispherical silicon lens, allowing the THz beam to be focused onto the photoconductive gap of the receiver. This allows us to perform single point reflection spectroscopy of targets at a stand-off distance of 1 m.

As a demonstration of the performance (measured in rapid scan and collected in a time of 1/15 s), Figure 7 shows the measured power spectrum with a mirror placed at the focal plane. From Figure 7, we show that, at a stand-off distance of 1 m, a system dynamic range of >60 dB, and a spectral bandwidth extending from 0.1 THz to about 3.0 THz is demonstrated. Sharp absorption features can be seen in the THz spectrum shown in Figure 7, which correspond to atmospheric water vapor lines—these can be removed algorithmically from the spectrum. These absorption bands in the THz spectrum are occurring after traversing a total path length of 2.4 m through the atmosphere. It can be seen that there are numerous water attenuation windows through which the signal is relatively unattenuated.

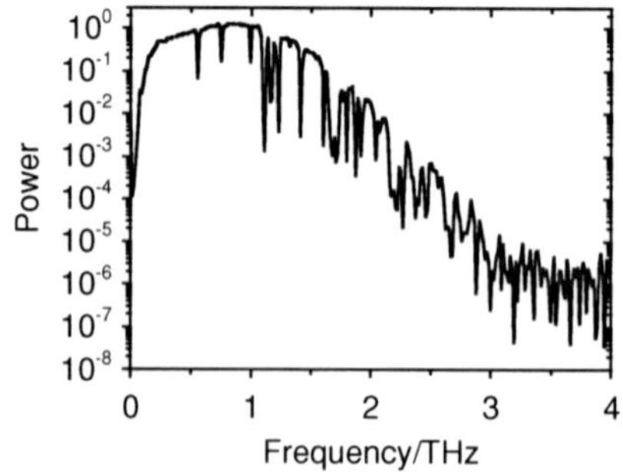

Figure 7. The power spectrum from the stand-off system with a mirror as the target material. The sharp absorption lines in the spectrum are due to atmospheric water vapour, with a THz path length of 2.4 m

In Figure 8, we show a reflection measurement from two explosive samples, (a) Semtex-H, and (b) SX2, taken using the stand-off instrument at a distance of 1 m, measured in rapid scan and collected in a time of 1/15 s. The absolute reflectance was determined by division of the measured power spectrum, by a power spectrum from a mirror reference, measured in an otherwise identical fashion (see Figure 7). Also shown in Figures 8(a) and (b) are the predicted reflection spectra of Semtex-H and SX2. The agreement between calculation and experiment is excellent, with features being observed in both spectra at 0.8 THz, 1.05 THz and 1.4 THz. For the measured spectra, data processing techniques were applied real time to the dataset to remove the water vapor absorption lines, i.e. both the data acquisition and data processing was performed in 1/15 s.

Also, there is agreement between the frequencies of the spectral features of Semtex-H and SX2. This is to be expected because both explosives contain

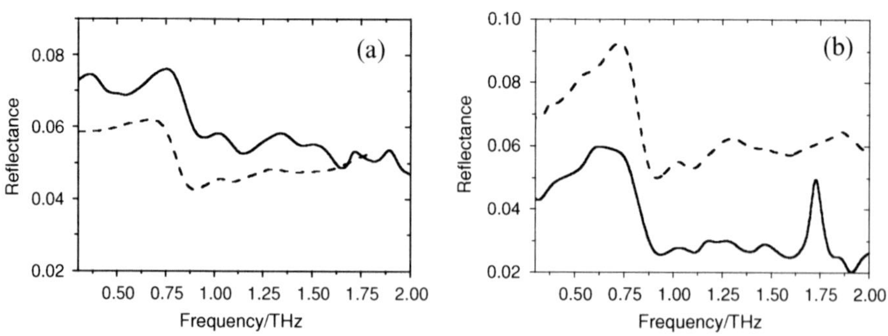

Figure 8. The measured (solid line) and calculated (dashed line) reflectance spectra of (a) Semtex-H, and (b) SX2 at a stand-off distance of 1 m, under normal atmospheric conditions, taken real time. The calculated spectra were derived from transmission spectroscopy data

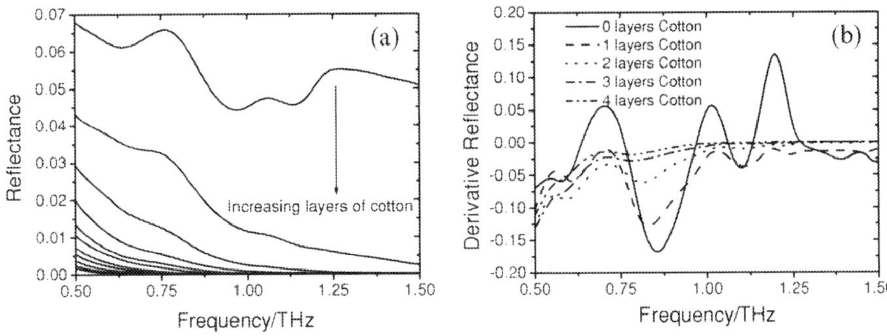

Figure 9. (a) The reflectance of SX2 behind cotton clothing, and (b) the first derivative of the reflectance of SX2 behind cotton clothing. The RDX spectral feature is clearly visible in (b) through 4 layers of cotton

substantial components of RDX, which dominate the spectral features over the frequency range of the figure. Since SX2 is predominately RDX (over 80% by volume), whereas Semtex-H contains only around 40%, the spectral features in SX2 are stronger than those from Semtex-H.

Clearly detection of explosives concealed behind clothing is crucial to the prospect of practical suicide bomber detection. We have performed a series of measurements of explosives hidden behind increasing numbers of clothing layers. Shown in Figure 7(a) is the effect of increasing numbers of clothing layers on the 1 m spectrum of SX2 sheet explosive. The clothing layers were from a typical cotton/polyester shirt loosely placed in front of the explosives, in a manner consistent with normal wear. As before these measurements were taken at a distance of 1 m, in a time of 1/15 s, and with automated water vapor removal. The rapidly diminishing reflectance, as the frequency rises, is a straightforward consequence of the exponentially increasing attenuation of clothing layers with frequency—the clothing layers acting as a low-pass filter to the spectrum. The effects of this can be visually removed by taking the derivative the reflectance spectrum, as shown in Figure 9(b). The reflectance feature of SX2 centered at 0.8 THz manifests itself as an oscillation in the derivative spectrum. Although difficult to recognize from the linear scaling of Figure 9(a), the derivative spectra (Figure 9(b)) clearly shows that we can convincingly identify spectral features through several layers of clothing.

As we have discussed in this paper, there are characteristic spectral features in many common types of explosives materials at THz frequencies. Relatively few other materials have strong spectral features in the range of 0.5–1.0 THz. A substance that does exhibit interesting features in this range is D-tartaric acid, as shown in Figure 10. This measurement was performed in reflection at a stand-off distance of 1 m, and was acquired under normal atmospheric

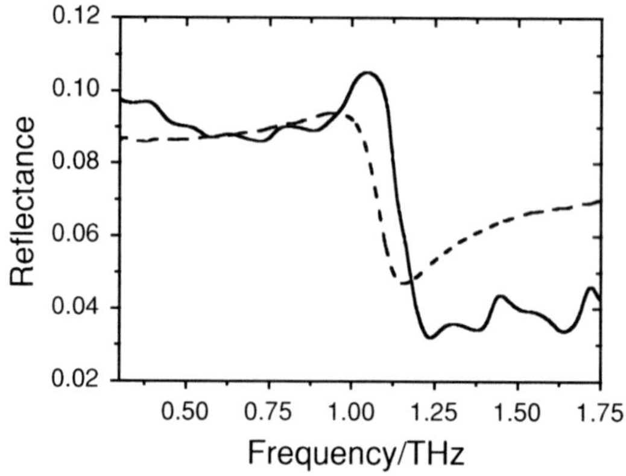

Figure 10. The measured (solid line) and calculated (dashed line) reflection spectrum of tartaric acid at a stand-off distance of 1 m. This demonstrates the ability of the technique to distinguish between materials

conditions, hence data processing algorithms to remove water vapor noise features. Both acquisition and data processing were performed in a time of 1/15 s. For comparison the calculated reflectance spectra of D-tartaric acid is shown as the dashed line in Figure 10, as determined from transmission spectroscopy. Again we can see an excellent agreement between calculation and experiment. Although the form of the reflection spectrum of D-tartaric acid is similar to that of RDX, the center frequency and magnitude is different, demonstrating that we can spectrally distinguish between explosives and other materials in reflection.

18.6. Future Directions

Whilst the results shown in the previous section are encouraging, they are not yet at a level where they can be used simply to construct an operational system. Further work is required in a number of areas:

- Increasing the operating distance and detectability through thick barriers.
- Extending the search area at the target from a single spot, to a scan over a whole person or a number of people.
- Meeting size, robustness and power-consumption requirements of an operational system.

Increasing the operating distance and barrier thickness principally requires continued improvements in system dynamic range. Algorithm and software improvements will also help meet this goal through, for example, better techniques for removing the effects of the (known) water vapour absorption lines from the atmosphere and improved pattern-matching algorithms for spectroscopic identification.

Extending the search area from the single-spot remote spectrometer model discussed here towards a system which can scan a crowd could be approached in two ways. First is to extend the terahertz capability through scanning or the use of detector arrays to produce a so-called "spectroscopic camera". Second is to combine THz spectroscopy with a more established imaging modality such as millimetre wave imaging. Here millimetre wave imaging could be used to identify suspect persons/regions and then THz spectroscopy is used as a tool to determine the nature of the threat.

Operational requirements may cause changes to the system architecture and particular terahertz technology employed in a fielded system. Pulsed terahertz systems as described here are currently the most mature THz technology. However, they are dependent on femtosecond pulsed lasers, which with their associated closed-cycle water chillers and power supplies are relatively bulky, fragile and expensive, and thus not ideal for portable THz systems for field deployment.

An alternative approach to THz systems is optical heterodyne conversion, or photomixing, which can be achieved using two continuous-wave (cw) lasers. The mixing of two above-bandgap (visible or near-infrared) wavelengths produces beating, which can modulate the conductance of a photoconductive switch (semiconductor) at the THz difference frequency. Upon application of a bias, monochromatic continuous-wave THz (cw-THz) radiation is produced. Furthermore, coherent homodyne detection is possible in the reverse scheme, in analogy with the optical gating mechanism for photoconductive detection of pulsed THz radiation. Since these all-photoconductive systems can be driven by inexpensive, compact and tunable diode lasers they have significant potential for low-cost, compact and robust THz systems.

Whilst this technology is less mature than pulsed terahertz systems, good progress is being made in its development and we have demonstrated imaging as well as rudimentary spectroscopy based both on swept frequency and simultaneous multi-frequency operation. Much of the work which is being carried out on optimizing semiconductor materials for pulsed terahertz systems can also be used to improve photomixer-based systems. The technology is also suited to the development of arrays—either as a means of increasing power and sensitivity, or as a means of rapid image capture.

18.7. Conclusions

In this paper, we have discussed some of the important factors involved in addressing practical THz systems for use in suicide bomber detection, and have demonstrated a prototype proof-of-principle terahertz stand-off detection system. Using this apparatus we have verified spectroscopic detection of explosives at a stand-off distance of 1 m, both real time, in reflection, under normal atmospheric conditions, and behind several layers of clothing.

These systems will become more sophisticated and sensitive if the software and technology is further developed and we have indicated a number of directions for further development towards operational systems.

Acknowledgements

Parts of this work were carried out under contract for the UK Government whose support is gratefully acknowledged.

References

[1] Auston, D.H., 1975. Picosecond optoelectronic switching and gating in silicon, Appl. Phys. Lett., 26, 101–103.

[2] Uhd Jepsen, P., R.H. Jacobsen, and S.R. Keiding, 1996. Generation and detection of terahertz pulses from biased semiconductor antennas, J. Opt. Soc. Am. B, 13, 2424–2436.

[3] Kemp, M.C., P.F. Taday, B.E. Cole, J.A. Cluff, A.J. Fitzgerald, and W.R. Tribe, 2003. Security applications of terahertz technology, SPIE 5070, 44.

[4] Tribe, W.R., D.A. Newnham, P.F. Taday, and M.C. Kemp, 2004. Hidden object detection: security applications of terahertz technology, SPIE 5354, 168.

[5] Strachan, C.J., P.F. Taday, D.A. Newnham, K.C. Gordon, J.A. Zeitler, M. Pepper, and T. Rades, 2005. Using terahertz pulsed spectroscopy to quantify pharmaceutical polymorphism and crystallinity, J. Pharma. Sci., 94, 837–846.

[6] Mittleman, D.M., R.H. Jacobsen, and M.C. Nuss, 1996. T-ray imaging IEEE J. Sel. Top. Quantum Electron., 2, 679–692.

[7] Pickwell, E., B.E. Cole, A.J. Fizgerald, M. Pepper, and V.P. Wallace, 2004. In vivo study of human skin using pulsed terahertz radiation, Phys. Med. Biol., 49, 1595–1607.

[8] Wallace, V.P., A.J. Fizgerald, S. Shankar, N. Flanagan, R. Pye, J. Cluff, and D.D. Arnone, 2004. Terahertz pulsed imaging of basal cell carcinoma ex vivo and in vivo, Brit. J. Dermatol., 151, 424–432.

[9] Smith P.R., D.H. Auston, and M.C. Nuss, 1988. Subpicosecond photoconducting dipole antennas, IEEE J. Quantum Electron., 24, 255–260.

[10] Piao, Z., M. Tani, and K. Sakai, 2000. Carrier dynamics and terahertz radiation in photoconductive antennas, Jpn. J. Appl. Phys., 39 (Part 1), 96–100.

[11] Smith, F.W., H.Q. Le, V. Diadiuk, M.A. Hollis, A.R. Calawa, S. Gupta, M. Frankel, D.R. Dykaar, G.A. Mourou, and T.Y. Hsiang, 1989. Picosecond GaAs-based photoconductive optoelectronic detectors, Appl. Phys. Lett., 54, 890–892.

[12] Chen, Y., S. Williamson, T. Brock, F.W. Smith, and A.R. Calawa, 1991. 375-GHz-bandwidth photoconductive detector, Appl. Phys. Lett., 59, 1984–1986.

[13] Tani M., K. Sakai, and H. Mimura, 1997. Ultrafast photoconductive detectors based on semi-insulating GaAs and In P, Jpn. J. Appl. Phys., 36, L1175–L1178.

[14] Kono, S., M. Tani, P. Gu, and K. Sakai, 2000. Detection of up to 20 THz with a low-temperature-grown GaAs photoconductive antenna gated with 15 fs light pulses, Appl. Phys. Lett., 77, 4104–4106.

[15] Baker C., I.S. Gregory, W.R. Tribe, I.V. Bradley, M.J. Evans, M. Withers, P.F. Taday, V.P. Wallace, E.H. Linfield, A.G. Davies, and M. Missous, 2003. Terahertz pulsed imaging with 1.06 μm laser excitation, Appl. Phys. Lett., 83, 4113–4115.

[16] Gregory, I.S., C. Baker, W.R. Tribe, M.J. Evans, H.E. Beere, E.H. Linfield, A.G. Davies, and M. Missous, 2003. High resistivity annealed low-temperature GaAs with 100 fs lifetimes, Appl. Phys. Lett., 83, 4199–4201.

Printed in the United States
59376LVS00001B/247-264